FASCINATING MATHEMATICAL People

FASCINATING MATHEMATICAL PEOPLE

INTERVIEWS AND MEMOIRS

**Donald J. Albers and
Gerald L. Alexanderson, Editors**

With a foreword by Philip J. Davis

PRINCETON UNIVERSITY PRESS • PRINCETON AND OXFORD

Copyright © 2011 by Princeton University Press

Published by Princeton University Press, 41 William
Street, Princeton, New Jersey 08540

In the United Kingdom: Princeton University Press,
6 Oxford Street, Woodstock, Oxfordshire OX20 1TW

press.princeton.edu

Library of Congress Cataloging-in-Publication Data

Fascinating mathematical people : interviews and
memoirs / Donald J. Albers and Gerald L. Alexanderson, editors ; with a foreword by Philip J. Davis.

     p. cm.

  Includes bibliographical references and index.

  ISBN 978-0-691-14829-8 (hardcover : alk. paper)

1. Mathematicians—Biography.   2. Mathematicians—
Interviews.   I. Albers, Donald J., 1941–
II. Alexanderson, Gerald L.

  QA28.F37 2011

  510.92'2—dc22

                    2011010624

British Library Cataloguing-in-Publication Data is
available

This book has been composed in Minion Pro and Eras

Printed on acid-free paper. ∞

Printed in the United States of America

10 9 8 7 6 5 4 3 2 1

Contents

Foreword

Philip J. Davis

How do the words of mathematicians, discussing their work, their careers, their lives, become known to a larger audience? There are, of course, biographies and autobiographies of mathematicians going as far back as Pythagoras. There are letters galore. Some off-the-cuff remarks have been preserved (e.g., those of Lagrange). Thus, authentic words of bygone mathematicians are not difficult to come by, and out of them it would be easy to construct an imaginative mock interview:

> *Interviewer*: Academician Euler, with so many children how did you manage to separate your professional life from your family life?
>
> *Euler*: I'm glad you asked. I didn't. I consider all of them, theorems and children, as my results.

Interviews? The word interview originally meant simply the meeting of two people. Later it came to mean the questioning of some notable person by a newspaper reporter. In the current and wider sense of the word, interviews with mathematicians start, say, around 1950. We now have interviews that link mathematical works, lives, and opinions. They are often called oral histories.

Since 1950 modes of communication and the possibility of social/technological interchanges have grown exponentially (a mathematical term that pundits like to use). Newspapers, books (e.g., the recent *Recountings: Conversations with MIT Mathematicians*), radio, TV, dramatic stage productions, movies, e-mail, YouTube, the Internet, and conference reports all record or depict interviews in which mathematicians often sound off on all sorts of topics far removed from their narrow specialties. I myself have interviewed several dozen numerical analysts, with texts now available online, as a part of a Society for Industrial and Applied Mathematics (SIAM) project. Thus, the time of much singing has come, and the voice of the mathematician is heard in the land. The question of who is listening is not beyond conjecture.

The variety of subjects, professional values, personal lifestyles, and more, of contemporary mathematicians are all well depicted in the interviews presented in this third collection of *Mathematical People*. Those of us in the mathematical business will find the macro aspect familiar, while the micro aspect satisfies our natural curiosity as to why and what the next fellow is up to.

Perhaps, therefore, the larger value of such collections as this may be found in, say, twenty-five or fifty years. During this interval of time, the very nature of mathematics, its contents, its applications, the manner of its presentation, and of its support by society may have changed substantially. The study of the history of mathematics confirms such a view. A future mathematician may pick up this book in a library or in whatever format texts are then the norm and say, "Gosh, that was certainly the golden age," or alternatively, and with considerable pride, "How much better, from all points of view, things are now!"

The best-known recent biographer of mathematicians is Constance Reid, the San Francisco writer, whose life of David Hilbert was highly praised when it was published in 1970. Her sister Julia Robinson was a distinguished University of California, Berkeley mathematician married to Raphael Robinson, also a Berkeley mathematician. Julia's marriage to Raphael and Berkeley's close proximity to San Francisco brought Constance in frequent contact with many mathematicians, and she found that she enjoyed their company and the culture in which they were embedded. Her interest in them eventually reached the point that she decided to write the Hilbert book, followed by volumes on Richard Courant, Jerzy Neyman, and E. T. Bell. After observing mathematicians for some time, she summarized her impression of them: "Mathematics is a world created by the mind of man, and mathematicians are people who devote their lives to what seems to me a wonderful kind of play!" It's true—if you're around a group of mathematicians excitedly talking about mathematics, you come away with a sense that whatever they're doing, it's fun.

Martin Gardner, the most widely read expositor of mathematics of the twentieth century, during the twenty-five years that he wrote a monthly column on mathematics in the *Scientific American,* came to know hundreds of mathematicians, many of whom had been inspired by his work. He corresponded with some of the best mathematicians in the world, who regularly sent him news of mathematical developments that he often fashioned into wonderful columns that were enjoyed by thousands of readers every month. One of us interviewed him a few years ago, and he had this to say about mathematicians: "There are some traits all mathematicians share. An obvious one is a sense of amazement over the infinite depth and the mysterious beauty and usefulness of mathematics."

Both of us have interacted with large numbers of mathematicians in our professional lives as professors, journal editors, and officers of the Mathematical Association of America. We have interviewed dozens of mathematicians over the last thirty years, and our conversations with and personal observations of scores of mathematicians lead us to strongly agree with the descriptions of mathematicians provided by Reid and Gardner.

Historically mathematics has often been motivated by questions in astronomy, mechanics, the physical and biological sciences, and, more recently, in economics. It is therefore widely appreciated because it works. It tells us about phenomena that are otherwise hard to explain and yet are important to society. Modern, so-called abstract, mathematics is not so easy to justify on the grounds of utility because much of it is developed from within the discipline of mathematics itself, just because it is interesting. But then the miracle occurs—suddenly recondite properties of seemingly useless prime numbers, for example, become essential as your ATM machine communicates safely with its central bank. Modern-day mathematics is certainly an example of the remark of the eminent physicist Eugene Wigner on "the unreasonable effectiveness of mathematics." Some abstract forms of mathematics seem far removed from practical scientific questions, and then sometimes we find these ideas popping up in applications when we least expect them.

Who then are these people who devote their lives to mathematics? Like other people, they come in a variety of shapes, sizes, and temperaments. Some are gregarious and outgoing, others reserved. In these interviews we see a wide range of people drawn to this beautiful subject. They come from diverse family backgrounds: blue collar, white collar, the socially prominent, or from long-standing academic families. Some are seriously religious, others are not, at least outwardly. Some were thinking about mathematical ideas in the early grades, and others did not come to the subject until high school, college, or graduate school. As you read these interviews, profiles, and memoirs you will come to the conclusion that there is no such thing as a typical mathematician. But all of them share a passion for mathematics (and, in many cases, other fields as well).

One common myth about mathematicians is that they are especially interested in music. It is true for some but not for others, although mathematics and music both deal with patterns. Scanning the mathematicians in our three volumes of *Mathematical People* interviews, we find no evidence of a strong music-mathematics connection. Mathematics is one of a very short list of disciplines where we find some child prodigies. We don't see historians who are child prodigies. To write good history one needs a lifetime of experience, something unnecessary in observing some deep pattern in the whole numbers, for example. Of course, not all successful mathematicians were prodigies, either.

In reading these interviews and profiles, the reader can perhaps recognize some patterns. But the diversity is probably the more interesting observation about their lives. Diversity is also evident in the range of talents of mathematicians within their subject.

Some are great problem solvers. Others are less talented at solving concrete problems but see vast structures within mathematics. Both talents are valuable. These structural questions are rewarding in themselves. For example, sometimes we see an enormously rich structure come from a minimal investment. Four seemingly straightforward and simple properties define the idea of a mathematical *group*, but they lead to an amazingly subtle and deep branch of mathematics that solved not only classical problems but a couple of centuries after their formulation still provide a way of approaching difficult problems in the sciences. One sees the same phenomenon in the seemingly small investment in the definition of an *analytic function* and then beholds the most amazing developments of the field of *complex analysis*. Small wonder that the subject is described as beautiful!

Does exceptional mathematical talent run in families? There's some evidence to suggest that it does. In these pages we read reminiscences of Alice Beckenbach whose father and brother were professional mathematicians of note. She married A. W. Tucker, the topologist and long-time head of the Mathematics Department at Princeton. She has two sons who are professional mathematicians, one at Colgate, the other at the State University of New York, Stony Brook. One of those, in turn, had a son, now a mathematician at the University of Rochester, and a daughter, who is a biologist now at the National University of Singapore. In that family there clearly were some "mathematical genes." The most renowned mathematical family in history was the Bernoulli clan, a Flemish family living in Basel: in addition to the obvious progenitors, Jacques, Jean, and Daniel, there were Bernoullis active in mathematics through three generations. They are the mathematical counterparts of the phenomenal musicians, the Bach family, in Eisenach and Leipzig. The Bernoullis do not make it into the current collection, nor do some more recent father-and-son combinations: Elie and Henri Cartan, George David and Garrett Birkhoff, as well as the earlier Benjamin and Charles Sanders Peirce.

Some here—Ahlfors, Cartwright, McDuff, and Selberg—are in the very top ranks of research mathematicians of the twentieth century, invited to be major speakers at International Congresses and, in two cases, winners of Fields Medals, the equivalent in mathematics, in some ways, of Nobel Prizes. Others are major figures in mathematics who have extensive publication records—Apostol, Banchoff, and Taylor—but have also acted as mentors and dissertation advisors, thus having had an influence on research mathematics far beyond their own campuses.

Then there are the extraordinary teachers, a list including some of those above but also others who have fine research records but who have had broad influence on twentieth-century science and mathematics: Bacon, Benjamin, and Gallian. Who is to say whose overall influence on the future directions of mathematics and science will be greatest?

There are also present here people working around the edges of traditionally central areas of mathematics: Saari (physics and astronomy, as well as, in later years, economics and social sciences) and Tondeur (public policy and mathematical administration).

It is not easy, however, to categorize these people because in large part they often excel in various activities, not just research or not just in teaching or administration. There is Bankoff, who probably could have had a good career in mathematics but chose instead to be a dentist, who just happened to edit a problem section of a mathematical journal and wrote mathematical articles on the side. He was the dentist of choice for many Hollywood stars and rock musicians. He was also a jazz pianist.

Gallian, a great teacher, is also an international expert on the Beatles. Benjamin enjoys a second career as a professional magician who has performed on a number of network television shows and on a host of national stages. Guy and Taylor are highly accomplished mountain climbers.

We are certain as you read this volume that you will encounter several mathematical people whom you would like to know better and with whom you would enjoy having dinner.

Donald J. Albers
Gerald L. Alexanderson
January 2011

Acknowledgments

We are grateful to all the individuals named below for so generously taking part in the making of this book.

All but four of the interviews and the one profile were done by Donald J. Albers and/or Gerald L. Alexanderson. Alice Beckenbach contributed a memoir. Her sons Alan and Tom Tucker were instrumental in seeing the memoir to publication. Claudia Henrion did the interview of Fern Hunt, which was originally published in *Women in Mathematics: The Addition of Difference* (Indiana University Press, 1997). Deanna Haunsperger did the interviews of Joe Gallian and Donald G. Saari, and James Tattersall and Shawnee McMurran did the interview of Dame Mary Cartwright.

Some of the interviews were first published in the *College Mathematics Journal*, a publication of the Mathematical Association of America (MAA). Thanks to the MAA for permission to reprint the following interviews: Ahlfors (March 1998), Apostol (Sept. 1997), Bankoff (March 1992), Cartwright (Sept. 2003), Gallian (May 2000), Guy (March 1993), Saari (March 2005), and Taylor (Sept. 1996).

Several people provided valuable assistance in making possible the publication of the Selberg interview: Betty Compton Selberg, Ingrid Selberg, Lars Selberg, Brian Conrey, Dennis Hejhal, and Peter Sarnak.

Thanks to Fernando Gouvea and Richard A. Scott for their work on the glossary. Leonard Klosinski, Mary Jackson, Geri Albers, and Jean Pedersen provided general assistance as the many months of manuscript preparation went by.

We are also grateful to the following individuals and organizations for permission to reproduce copyrighted photos and figures, referenced below by figure number. Abbreviations used are American Mathematical Society (AMS), Mathematical Association of America (MAA), Mathematisches Forschungsinstitut Oberwolfach (MFO), Donald J. Albers Collection (DJA), and Gerald L. Alexanderson Collection (GLA).

Figures 1.1–1.13, except 1.2, 1.3, and 1.7: Caroline Ahlfors Mouris; 1.2: mikkeli.fi; 1.3: GLA; 1.7: DJA; 2.1–2.14, except 2.7 and 2.8: Tom Apostol; 2.7: MFO; 2.8: GLA; 3.1–3.9, except 3.2 and 3.7: Charles Bacon; 3.2 and 3.7: GLA; 4.1–4.15, except 4.5: Tom Banchoff; 4.5: Stephen Simon; 5.1–5.12: Janis Bankoff; 6.1 and 6.2: GLA; 6.3 and 6.4: DJA; 6.5: MAA; 6.6: DJA; 6.7: GLA; 6.8: DJA; 6.9: AMS; 6.10: GLA; 6.11: AMS; 6.12: MAA; 6.13: GLA; 6.14: DJA; 6.15: Institute for Advanced Study Archives; 6.16 and 6.17: DJA; 7.1–7.9: Arthur Benjamin; 8.1 and 8.2: Girton College Library and Archive; 8.3: GLA; 8.4 and

8.5: Harold P. Boas; 8.6: South Wales Evening Post; 8.7: GLA; 8.8: Cambridge University Press; 8.9: DJA; 8.10: Girton College Library and Archive; 9.1–9.9: Joe Gallian; 10.1–10.19: Richard K. Guy; 11.1–11.6: Fern Hunt; 12.1–12.16: Dusa McDuff; 13.1–13.9: Donald G. Saari; 14.1: Betty Compton Selberg; 14.2: C. J. Mozzochi; 14.3: Lars Selberg; 14.4: Betty Compton Selberg; 14.5: Nils Baas; 14.6: C. J. Mozzochi; 14.7: Yangbo Ye; 14.8: Nils Baas; 14.9: Ingrid Selberg; 14.10: Yangbo Ye; 15.1–15.9: Jean Taylor; 16.1–16.11, except 16.4 and 16.10: Philippe Tondeur; 16.4 and 16.10: MFO.

Working with the staff of Princeton University Press has been a pleasure. Vickie Kearn, the mathematics editor, has been wonderfully supportive throughout the entire publication process, from book idea to a finished volume. Her able assistant Stefani Wexler gave valuable help throughout the production phase, as did Elissa Schiff, our copyeditor. The published book demonstrates the considerable production talents of Natalie Baan. The design ability of Lorraine Doneker shines through the entire volume.

To all of you who worked on this book, we again say thank you.

Sources

Lars V. Ahlfors: *College Mathematics Journal* 29:2 (1998), 82–92. Interview conducted at the Ahlfors family cottage at East Booth Bay, Maine, in June 1991. Also present were Ahlfors's wife and daughter.

Tom Apostol: *College Mathematics Journal* 28:4 (1997), 152–70. Interview conducted on the Caltech campus in 1993 and concluded in 1997.

Tom Banchoff: Interview conducted in Louisville, Kentucky, at the Joint Mathematics Meetings in 1990, continued and updated at the Institute for Advanced Study, August 1999.

Leon Bankoff: *College Mathematics Journal* 23:2 (1992), 98–112. Interview conducted in Santa Clara, California, August 22, 1986 and continued in San Francisco in January 1991 at the Joint Mathematics Meetings where Bankoff was speaking at a special session on geometry.

Arthur Benjamin: Interview conducted July 17, 1998, at the Toronto Mathfest; updated and expanded in August 2009.

Dame Mary L. Cartwright: *College Mathematics Journal* 32:4 (2001), 242–54. Interview conducted in person and through correspondence during the period 1990–94.

Joe Gallian: *College Mathematics Journal* 39:3 (2008), 174–90. Interview conducted in Washington, D.C., May 2007.

Richard K. Guy: *College Mathematics Journal* 24:2 (1993), 122–48. Interview conducted in Menlo Park, California, in the spring of 1988.

Fern Hunt: *Women in Mathematics: The Addition of Difference* by Claudia Henrion, Bloomington and Indianapolis, Indiana University Press, 1997, pp. 212–33.

Dusa McDuff: Interview conducted at the Mathematical Sciences Research Institute (MSRI) in Berkeley, California, April and May 2010.

Donald G. Saari: *College Mathematics Journal* 36:2 (2005), 90–100. Interview conducted at the Boulder Mathfest of the Mathematical Association of America in 2003.

Atle Selberg: Interview conducted in Selberg's office at the American Institute of Mathematics in Palo Alto, California, June 1999.

Jean Taylor: *College Mathematics Journal* 27:4 (1996), 250–66. Interview conducted in Taylor's office at Rutgers University in the summer of 1994.

Philippe Tondeur: Interview conducted in the offices of the Mathematical Association of America in Washington, D.C., in the fall of 2002 and continued in 2005, November 2007, and January 2008.

FASCINATING MATHEMATICAL People

One

Lars V. Ahlfors

Donald J. Albers

Lars Valerian Ahlfors (1907–1996) was born in Helsinki, Finland, and earned his PhD at the University of Helsinki under Ernst Lindelöf in 1928. In 1935 he came to Harvard University for three years but then returned to Helsinki to help carry on its mathematical tradition. Forced to leave by World War II, he accepted a professorship at Harvard, where he worked until his retirement in 1977. Ahlfors is remembered by his students and colleagues as refined and reserved, yet a delightful host fond of good food, drink, and conversation—and always eager to work the following day.

Ahlfors's work virtually defined the subject of complex function theory in the middle half of the twentieth century. His textbook *Complex Analysis* has attracted generations of students to the subject. In his research papers he introduced and developed the theory of quasiconformal mappings, for which he received a Fields Medal in 1936. In collaboration with Lipman Bers he made pioneering advances in the theory of Kleinian groups, and Ahlfors's contributions to geometric function theory underlie much of modern complex manifold theory. The breadth and depth of his work are shown by his being the only

mathematician ever invited to give three plenary addresses to International Congresses of Mathematicians.

The Boy Who Loved Homework

MP: *Ahlfors is a Swedish name. Did your parents come from Sweden and settle in Finland?*

Ahlfors: Oh, no! My father was born on the Åland Islands, which is a part of Finland but has a Swedish population and is completely Swedish-speaking. My ancestors came to Mariehamm in the Åland Islands from Björneborg, a smaller city on the mainland. I think that some of them were sea captains with sailing ships that traded all over the world. My grandfather had a grocery store, and he must have done fairly well because he was able to send his children to school on the mainland. This was good since there were practically no schools on the Åland Islands. That's how my father was able to become first an engineer and then a professor of engineering at the Polytechnical Institute in Helsinki.

MP: *What do you remember about your mother?*

Ahlfors: I didn't know her at all. She died in childbirth. When I was a newborn, I was sent to the Åland Islands to be taken care of by two aunts. At two or three I was sent to my father in Helsinki. At that time a professor's salary was excellent, and we had lots of money. I remember the days when we had a maid, a cook, and a governess. Who can do that on a professor's salary today? On the other hand, the domestics were paid very little.

MP: *What was your father like? As a professor of mechanical engineering, he obviously had some mathematical ability.*

Ahlfors: He was very stern. Yes, he did have some mathematical ability. When I was about twelve or thirteen, I would make clandestine visits to his engineering library at home, to find mathematical books.

MP: *When did mathematics become particularly interesting to you?*

Ahlfors: As a young boy. When I was three or four, my father started to ask me

Figure 1.1 Ahlfors (right) and Trols Jorgensen on the lawn of his Maine cottage.

mathematical questions, and I had to give the answers quickly. So it was quite obvious that at an early age I had a special sense of mathematics. I was fortunate in having two sisters who were two and three years older. I followed their mathematical education, and I loved to do their homework. So that's the way I slowly began to know some mathematics.

On the other hand, I was not given any books on mathematics. My father did not press me in any way at all to move ahead of what I was supposed to do in class. Unfortunately, the level of mathematics in my school and even in high school was not very advanced. No calculus was taught. In fact, the teachers used me to make up new problems.

MP: *Was that true throughout Finland?*

Ahlfors: No. Some schools offered advanced mathematics, but not the one that my father sent me to. It was a private school that was supposed to be better in some ways—in particular, it was extremely good in languages. When I left Finland for the first time and traveled to Berlin, I could understand everything. I had an excellent sense of grammar, so I could speak correct German fluently straight out of school.

MP: *You studied languages over a period of four years?*

Ahlfors: Something like that. You see, first I went to a Swedish-speaking school, but everybody had to study Finnish from the start. So quite young I had at least an idea of Finnish—although that is not enough to learn Finnish!

MP: *Were sports an interest of yours in school?*

Ahlfors: None whatsoever! I hated sports. I also hated vacations and Sundays, for I had nothing to do on those days.

MP: *Many Scandinavians love to go up to the mountains.*

Ahlfors: Not I. I loved what you would call the homework exercises in my school courses. I also had access to books. I was not taught calculus, but I learned it from my father's engineering books.

MP: *Apart from mathematics, were there subjects that particularly interested you?*

Ahlfors: I worked hard on languages. I did not like history. Why memorize the years associated with various events and people? One might just as well memorize telephone numbers. It didn't make any sense to me, and it seemed rather silly. My history teacher was not very fond of me.

MP: *Given your interests, and your father's prompting, was it almost preordained that you would end up in mathematics?*

Ahlfors: Oh, yes, I think so. It became absolutely clear to me one day when I was about fifteen, as I was walking in Helsinki with my father. We met the only full professor of mathematics at Helsinki University, and he and my father politely took off their hats as we passed. Then, using my family nickname, my father said to me, "That is what you should become, Lasse. You shall become a professor of mathematics."

MP: *Your father must have been quite aware of your leanings.*

Ahlfors: Yes, he was. He had wanted me to become an engineer, like himself, but had soon found out that I could not do anything mechanical, not even put in a screw. So he decided that I should go into mathematics.

Discovering Real Mathematics

MP: *The mathematician whom you met on your walk, was that Ernst Lindelöf?*

Ahlfors: Yes, Lindelöf—the father of mathematics in Finland. In the 1920s all Finnish mathematicians were his students. He was essentially self-taught.

Figure 1.2 Ernst Lindelöf was the thesis advisor of both Ahlfors and Nevanlinna. Ahlfors dedicated his classic text, *Complex Analysis*, to him.

MP: *Did he teach you?*

Ahlfors: I was very lucky when I finally came to the University, for I started out with two great teachers, Lindelöf and Rolf Nevanlinna. I don't believe I've had better teachers. I was seventeen when I entered the University, and I had carefully read all the information for new students, which said not to be afraid to ask the professors questions. So I'm afraid that I asked a lot of questions! I have had students of my own who abused this privilege. But I was the best as far as the professors were concerned, and they were very good to me.

I took advanced calculus as a freshman, which I was not supposed to do. On the first day of school, when I went to Nevanlinna to ask permission to take his course, he asked me, "Do you know any calculus?" "Oh, yes," I said, "I have studied calculus." I didn't tell him it was all on my own. Lindelöf did not like my jumping ahead, and he made me take all of the tests from the prerequisite courses that I had not attended. I liked advanced calculus; I was very interested in my work.

MP: *The original calculus that you learned must have been computationally oriented, so you essentially learned techniques. Was that in fact helpful? Some people believe that when it comes to looking at analysis and really understanding the theory, it's a great benefit to know the mechanics.*

Ahlfors: I would not say that it is bad to start with the mechanics of calculus. On the contrary, one must have a lot of technique even to learn calculus.

MP: *During those years with Lindelöf and Nevanlinna, how much complex analysis did you get?*

Ahlfors: Well, it was not possible to have a full set of advanced courses each year at Helsinki because there were too few professors. So one course was given one year and another the next year, and so on. Thus, one depended very much on what was offered, but because of Lindelöf, complex analysis had already become a special subject for the Finnish mathematicians. I began studying it in my second year.

During my first year, Lindelöf told me that something strange had happened: Nevanlinna had been invited to Zürich to substitute for Hermann Weyl, who had gone to America for a year. Lindelöf said I must try to go to Zürich to hear Nevanlinna lecture. So that's what I did!

Zürich was my first encounter with mathematics as it should be taught. It was contemporary mathematics, not the history of mathematics as our courses usually were. In Zürich Nevanlinna could talk on the research level, which made a completely different impression on me. I came to understand what mathematics was about and that I was supposed to do mathematics, not just learn it. This had not been clear to me before.

MP: *Was Pólya at Zürich when you were there?*

Ahlfors: Yes. Pólya was perhaps the most interesting person at Zürich. He led the seminar, which of course was very good—but he had strange ways of doing it.

Figure 1.3 Rolf Nevanlinna, a great influence on Ahlfors.

Hearing about unsolved problems is what really kept me going. Early in my stay, I had the good luck to solve the Denjoy conjecture, so I became a known mathematician right away. It was not so terribly important, but it was new, and it showed the way to other things.

MP: *You were still very young then.*

Ahlfors: I was twenty-one. I followed Nevanlinna to Paris for the spring term after my term in Zürich. That's why I really worked a lot on his theories. In Paris I also got to meet Denjoy. He told me that now he knew why twenty-one is the most beautiful number: because twenty-one years ago he had made this conjecture, and when I was twenty-one I had solved the problem.

Helsinki to Harvard

MP: *Most young men of twenty-one have few obligations—no wife, children, and so on. And having made a name for yourself, you were able to travel around and talk to people more easily. Was this why it took a few years to complete your dissertation?*

Ahlfors: Well, since I was on my own, I could sit up half the night and work. But my dissertation developed into a rather important paper because it used the same methods that I had used in the Denjoy conjecture, although a much broader aspect of them. Also, upon my return to Finland in the fall of 1929, I was made a lecturer at Aaboe Academy in Turku. It was like a small university, Swedish-speaking, and I taught there for three years. Since I did not care for Turku, I commuted to Helsinki.

MP: *Do you like to teach? Is it fun for you?*

Ahlfors: Yes, I like to teach. It's not the same as doing mathematics; you have to think about the subject from a different angle. For

Figure 1.4 A young Ahlfors with a look of determination.

Figure 1.5 Ahlfors was fond of berets.

one thing, you teach things that were done by others and that you have not done yourself. Also you must empathize—think about what the students really need to learn and how they should be taught. I'm not saying that I'm a terribly good teacher, but I certainly enjoy teaching.

MP: *Where did you go after Aaboe Academy?*

Ahlfors: The next step for me was a position at Helsinki University. Then I got a Rockefeller grant for an academic year in Paris, which was a great experience for me. Then in 1935 I went to Harvard.

MP: *How did you react to Harvard?*

Ahlfors: I liked it from the start—not only the mathematics but the whole atmosphere. It was excellent. The department was extremely friendly. It was like a big family, which my wife especially appreciated. The Depression was still on in America, but Harvard paid me $3000—an enormous amount of money at that time.

Mrs. Ahlfors: William Caspar Graustein, a Harvard mathematics professor, made $9000, and I asked him, "What do you do with so much money?"

Ahlfors: You could travel all over the world and rent golden Cadillacs!

MP: *Of course, Harvard was the leading American institution.*

Ahlfors: Yes. Many of the others were quite poor. But for years now, Harvard has not seen that being a leading institution is a reason to pay people well. They used to say Harvard stands for excellence, and you can't get excellence unless you pay for it. But then they decided that excellence was too expensive. Somehow it fell down quite a bit.

Mrs. Ahlfors: And of course we also had the "gravy train."

Ahlfors: Oh, yes; one could have additional gravy by repeating one's lectures for Radcliffe students. They were not allowed to hear lectures at Harvard, but the professors were allowed to give their lectures at Radcliffe and were paid $1000 extra per year. So in my second year, I was already at $4000.

MP: *When you arrived at Harvard, who was the chairman of the mathematics department?*

Ahlfors: Graustein. He told my wife that he had written to the noted mathematician Constantin Carathéodory for advice about whom to invite to Harvard, and Carathéodory wrote back: "There is a young Finn, Ahlfors, and you should try to get him."

MP: *Yet it must have been difficult to leave Finland.*

Ahlfors: Well, we were a little bit homesick. We agreed to a mutual probation period of three years, which was just about right. And then in 1938, when the three years were over, my teacher Lindelöf sent a letter saying that Finland can export goods, but not intelligence. He asked me to come home.

Figure 1.6 Ahlfors in lecture mode.

Figure 1.7 Constantin Carathéodory.

Papa in Wartime

MP: *In 1938, when you went home to Finland, Europe was already beginning to break apart.*

Ahlfors: We had one very, very happy year in Finland, but then in 1939 came the war. Yet I would not have wanted to be in America when Finland was at war.

MP: *With so many real dangers in most parts of Europe during the war, I am impressed that people stuck it out.*

Daughter: There was no place to go.

Ahlfors: Being a Finnish citizen, I was never in the military service, but I was still considered eligible for special service.

Daughter: Tell what you did during the war, Papa.

Ahlfors: Well, I answered the telephone.

Mrs. Ahlfors: But it never rang.

Ahlfors: No, it never rang! Actually, I was one of three mathematicians assigned on eight-hour shifts to watch a telephone twenty-four hours a day. We were in a shelter that couldn't easily be bombed, a rather safe place. If everything went wrong, then we were to be connected with military headquarters in order to code their messages. But nothing ever happened.

MP: *In 1945 you decided to return to Zürich. Was it difficult to leave Finland at that time?*

Ahlfors: Normally I would not have been allowed to leave at all, since I was a Finnish citizen. But my health was not in the best condition; I have an irregular heartbeat. My doctors said that I needed to go to Sweden, which was also good for my family.

MP: *So they were able to come with you. Did anyone realize that you were intending to leave Finland?*

Ahlfors: Everybody knew. Nevanlinna was one of the first to know it, and he encouraged me to leave because everyone realized that it would be a long time before conditions in Finland would again be normal so that one could do serious research.

MP: *How did you get out of Sweden?*

Ahlfors: Sweden was neutral in the war, but getting from Sweden to Switzerland was very hard. Finland had already made an armistice with Russia and was at war with everybody else—both the Allies and Germany. So the Germans would not let us fly over Germany, but the Russians would not let us fly over, either.

Although officially Finland was at war with England, the British decided to help us. There were American military planes that had been shot down in Sweden and then repaired and put back into service. They flew between Stockholm and an airfield in Scotland—but only on nights with no moon and at a high altitude, up in the stratosphere. We had to use oxygen masks all the time, which was not easy with small children. They also warned us that if we were shot down and landed on the water, we would have to get out on the wings.

After getting to England finally, we then had to cross the English Channel to France. But waiting for the boat-train took a long time. We would take everything to the train station, where we sat on the train, waiting for the boat. At last they would come and tell us that there was no boat that night, and we would go back to the hotel. This was repeated perhaps ten times, until finally one day the boat came and the train truly left. This was in early 1945. London, of course, was being bombed by the V-2s, and the children were very scared. When we crossed the English Channel, everybody had to put on life jackets. The children argued, "If there is no danger, why do we have to put them

on?" And then, on the train from Paris to Zürich, we were bombed by the Germans all the time.

MP: *You really wanted to get to Zürich.*

Ahlfors: We certainly wanted to get to Zürich, and the Swiss wanted us to come, too.

MP: *How did the British decide who was sufficiently important to take one of their special high-altitude flights from Sweden?*

Ahlfors: I think that the Swiss laid diplomatic groundwork for people all the time. In fact, they had even contacted hotels in London and in Paris to accommodate us when we arrived. In Paris, we stayed in the George V, which is the most fantastic hotel I have ever seen. It was quite inexpensive for us, because in those tumultuous times nobody knew what money was worth. I had British pounds, and no one in France knew how many francs to exchange.

Finally, we got to Zürich, and a year and a half later, I was asked to come back to Harvard. When Hermann Weyl heard that I was in Zürich, he understood that I no longer felt a patriotic reason to stay in Finland. So he let Harvard know that they should try to get me.

The First Fields Medal Winner

MP: *Back in 1936 you were still at Harvard, so they could claim you at the International Congress of Mathematicians in Oslo that year. That was the first time the Fields Medals were*

Figure 1.8 Ahlfors with his three daughters in the Austrian Alps in the early fifties.

Figure 1.9 Giving away his daughter Cynthia at her wedding in Paris.

awarded, and you were the first winner. Did you have any prior knowledge of this honor?

Ahlfors: I learned of it only a few hours before receiving it. I don't know what would have happened if I had decided not to go to Oslo! George Birkhoff of Harvard knew about the Fields Medals, so he probably would have done something about it if I had said that I was not going.

MP: *In 1936, a Fields Medal had no special significance yet.*

Ahlfors: No; people did not know what it was. Although I felt very honored, I thought

that there must have been somebody better than I who deserved it. Now I am not so sure: my competitors had not reached their heights then; they were not sufficiently well known.

MP: *What, if anything, has the Fields Medal meant to your career?*

Ahlfors: Oh, quite a bit—though not so much in the beginning, I must say. When I returned to Harvard I was hired as only an Assistant Professor. That would never happen anymore.

MP: *But Harvard was still very pleased that you won.*

Ahlfors: Oh, yes. I was very pleased, too. It certainly helped me when I wanted to get away from Finland because of the war.

MP: *Did the Fields Medal also make a positive contribution to your career later?*

Ahlfors: I did not take it too seriously at the time. Actually, I was not sure at all whether it was a good idea. But I think that it has turned out to be one.

MP: *Why do you think the Fields Medal is a good idea?*

Ahlfors: Because it keeps mathematics on a very high level. I really think that it spurs people to try hard. Possibly the same people would do the same mathematics whether or not the Fields Medal exists, but I think that letting people know that they are good can be quite important for how they work.

MP: *What do you think of comparisons with the Nobel Prize?*

Ahlfors: It is strange that it is sometimes referred to as the "mathematical Nobel Prize." It is not a *prize* because there is no money. I would have welcomed $100,000! I did get the Wolf Prize, and that was a good one—$50,000. And I got a prize in Finland that allowed me to buy my summer home in Maine.

It is probably a good thing that there is not a Nobel Prize in mathematics. Since they have to pick a winner every year, they would run out of good mathematicians, I'm sure.

Certainly for economics, which was not one of the original Nobel Prizes, they have to scrape the bottom of the barrel.

MP: *What other personal benefits have you had from winning the Fields Medal?*

Ahlfors: I can give one very definite benefit. When I was able to leave Finland and go to Sweden, I was not allowed to take more than 10 crowns with me. So what did I do? I smuggled out my Fields Medal, and I pawned it! I'm sure it is the only Fields Medal that has been in a pawn shop.

MP: *Did you get it back?*

Ahlfors: Oh, yes. As soon as I got a little money some people in Sweden helped me retrieve it.

Doing Mathematics

MP: *How do you do mathematics? Do you use geometric images? Your subject is certainly geometrical.*

Ahlfors: Well, the way I do it has changed over time. Of course, the nature of the problem changes. And it becomes harder and harder when one gets older; it's just not quite so easy anymore. I have always relied very much on the subconscious. One works, and works hard, but one does not really discover anything while working. It is later that one makes discoveries.

I still do mathematics in bed. When I wake up in the morning, I think much more clearly than at night. In the morning

Figure 1.10 Ahlfors taking a break at a mathematics meeting.

everything in my system runs better! Sometimes, of course, when one gets an idea, it's not very good; but one has to take the bad with the good.

MP: *Do you think in geometric images, or in terms of symbols?*

Ahlfors: It's mostly a logical process: as one thinks, one looks for the logical connection. Of course, to put that on paper is hard work. There are many mistakes; one learns from the mistakes, and so on. It's a complicated process.

MP: *Are mathematicians lonely?*

Ahlfors: I wouldn't say that. I think that mathematicians are very nice people on the

Figure 1.11 A cigar and a mathematical model—a delightful pairing.

whole. That's my experience. It's very nice to have mathematicians as friends.

MP: *How about physicists?*

Ahlfors: Well, I don't believe in physics!

MP: *You don't believe in physics? Why not?*

Ahlfors: Physicists are so close to mathematics, but they don't know mathematics.

MP: *Some of them do. How about John Wheeler, who lives just across the water from you? His work shows quite a bit of high-powered mathematics. There's also a great deal of mathematics used by string theorists.*

Figure 1.12 Ahlfors receiving an honorary degree from the English Queen Mother.

Ahlfors: But it's the wrong theory. I like the knot theory aspects, especially the knot theory applied to string theory. The strings are knots now, and there are these ready-made knot theorems that can be applied. That appeals to me.

Probably physicists are important for mathematics, but they cannot be important for me in any sense. I don't think that mathematicians should take their inspiration from physics.

MP: *You are now more than four times twenty-one. Do you ever worry about not being able to do mathematics anymore?*

Ahlfors: No. I know of mathematicians who no longer do mathematics because they are afraid that it would not compare with what they have done before. I am not afraid of that. I can see that I have a hard time, and I can see that I make mistakes. But I always find the mistakes and learn from them, so I am not too scared. I am still doing something that I think will be good—but, of course, I cannot say before it is finished. And it is true, it has taken me longer than I thought it would, but I am still confident that it will be good mathematics when it comes out.

MP: *Have there been periods in your life when you have been on a mathematical roll, when things were just popping along?*

Ahlfors: No. I can't say that. It doesn't come easy for me.

MP: *While you are in Maine during the summer, are you thinking good, pure mathematical*

Figure 1.13 Ahlfors with his wife Erna in 1987.

thoughts most of the time? Is mathematics just bubbling up around here for you?

Ahlfors: Yes; yes it is. I don't know whether it should be or not, but it is. Of course I cannot isolate myself completely and work as much as I would like to. There are other persons to take into account.

Don't Get Too Serious with a Mathematician!

Daughter: Growing up, we always were in awe of what Father did. We had no idea what it was. We still don't know. He tells us he's a philosopher.

Mrs. Ahlfors: It was sixty years ago when I got to know him. He had just come from Paris, and he said, "You know, you could be even more pretty if you would paint your face." In Paris the style was a chalk-white face with bright, red lips and black eyelashes. I started to put makeup on, and then I caught him. But it took me years to convince him that marriage would be a good idea.

MP: That's a long courtship even by standards of that time. How did you meet?

Mrs. Ahlfors: I was twenty-one years old, and he was working on his PhD dissertation. He was very young. We were introduced by a school friend of Lars's who was also a friend of mine. At the time, I was more interested in my cousin from Denmark, but when I met Lars I fell in love with his voice.

When I told my father that I had a new boyfriend, Father asked, "What does he do?" I said, "I don't know, but he is a mathematician." Father said, "Don't get too serious with him. You'll never be able to afford to get married."

MP: *Do you think the reaction would be much different today?*

Mrs. Ahlfors: Oh, yes. But at that time, it was not so. Father was very concerned. He warned me, "Don't get too serious with a mathematician!"

Ahlfors: Mathematics is a recognized profession now. At that time mathematicians were regarded as people in an ivory tower. Today, of course, the utilitarian aspect of mathematics plays a large role, and that is better known to society as a whole. Mathematics is becoming so much more important.

Two

Tom Apostol

Donald J. Albers

Tom M. Apostol is professor emeritus of mathematics at the California Institute of Technology. His writing talents blossomed in graduate school in the late forties at UC Berkeley, in large part because textbooks were rarely used by the UC faculty, and he felt compelled to transform his class notes into text form. He is especially well known for his pathbreaking *Calculus* in two volumes, first published in 1961, and his *Mathematical Analysis*, published in 1957. Both have been translated into several languages, and both are still in print! These books have had a strong influence on an entire generation of mathematicians. His other books include *Introduction to Analytic Number Theory*, *Modular Functions and Dirichlet Series in Number Theory*, *Selected Papers on Pre-calculus* (with Chakerian, Darden, and Neff), *The Mechanical Universe: Introduction to Mechanics and Heat* (with Olenick and Goodstein), *The Mechanical Universe: Mechanics and Heat for Science and Engineering Students* (with Frautschi, Olenick, and Goodstein), and *Beyond the Mechanical Universe* (with Olenick and Goodstein). In 1982 Apostol was initiated into the world of television production. Over the past ten years, he has developed a prizewinning series of videos titled *Project MATHEMATICS!* Although he has received considerable praise for his writing, he says that "of all the things I've done in my fifty years as a mathematician, the most satisfying has been producing and directing the videos for Project MATHEMATICS! . . . Making a video is like solving a puzzle."

The Immigrant Experience

MP: *Where did you start life?*

Apostol: I was born in Helper, Utah, in 1923. My mother, whose maiden name was Efrosini Pappathanasopoulos, was a mail-order bride from Greece. My father, Emmanouil Apostolopoulos, whose name was shortened to Mike Apostol when he became an American citizen, traveled from Utah to New York to meet her. She had been eager to come to the United States, but ending up in the mountains of Utah was not exactly what she had expected.

MP: *How long had your father been in America before your mother arrived?*

Apostol: My father came around 1916 and she came in 1922, so it was about six years.

MP: *What drew him here?*

Apostol: He had two brothers working in mine smelters in Montana, and they paid his way to come join them. They earned seventy-five cents a day. When his brothers had saved enough money they returned to Greece. But my father heard that the coal mines in Utah were paying a dollar and a quarter a day, and he thought, "Well, that's a great deal." So he went to Utah by himself in 1918, and he worked in a coal mine—but only for one day! When he climbed on the elevator and it descended 5,000 feet into the mine, he got claustrophobic. He just couldn't stand it.

MP: *But what else could he do?*

Apostol: The coal mine was at Castle Gate, about four miles north of a hamlet called Helper, the hub of Carbon County, and he saw an opportunity there. Since most of the miners had their shoes repaired in Price, seven miles south of Helper, he thought it would be a good idea to open a shoe shop in Helper. Back in Greece he had been apprenticed as a shoemaker from the age of eight, so he knew how to repair and to make shoes by hand. He had even brought a few hand tools with him. He rented a storefront, lived in the back room, and started repairing shoes.

MP: *Did his idea work out all right?*

Apostol: He put all the money he earned in a coffee can, and at the end of the first month he dumped it on the bed. There was $200—and for somebody who had been earning 75 cents a day, that was a small fortune. Most of it he sent back to Greece for his sisters' dowries. Then after all his sisters were married off, he decided it was his turn. Now, my father couldn't write very well, since he had only gone to the first or second grade, so he had a Greek friend write a letter to my mother's older brother in Egion asking for the hand of one of the daughters. Which one didn't matter, because he didn't really know them. He knew it was a good, respectable family with several daughters. They had a nice home with a lemon orchard and olive groves. My father's letter, which still exists, explained that he had paid the dowries of his sisters and wanted to start his own family now. So the letter arrived, but the eldest daughter, who normally would be the first to be married, didn't want to go to America by herself. My mother was the youngest daughter. She had heard great things about America, and she immediately spoke up and said, "I'll go." So her brothers put her on a steamship and off she went.

MP: *She must have had a lot of courage.*

Apostol: Yes, indeed. My father traveled across the country by train and met her in New York City. They were married promptly and took the train back to Utah. My mother told me that as they crossed the beautiful green fields of the Midwest she would say, "Oh, this looks like a nice place to live." And

Figure 2.1 Efrosini Pappathanasopoulos bidding farewell to her brother before she left Greece in 1922.

Corinth. She shed a lot of tears in the next few months.

MP: *How much younger than your father was your mother?*

Apostol: She was 22 when she came to America, six years younger than my father. He had bought a small house on Main Street, but it was only about 50 yards from the railroad tracks. Trains went by all day and night pumping out soot on her clean laundry. It was not really a happy time for her at first. And then I was born, the first child. She lavished all her attention on me, and the tears stopped flowing. All four of her

my father would say, "Oh, that's nothing. Wait until you see Utah."

MP: *So they landed in Helper, Utah. That's a pretty desolate place.*

Apostol: Right, about 5,500 feet elevation, in the middle of nowhere, not too many trees around—my mother was pretty upset when she got there. She had left this beautiful place in Greece overlooking the Gulf of

Figure 2.2 Apostol (age 14 months) with his mother.

children, two boys and two girls, were born in that house. Being the first child, and a boy, I got a lot of special attention. My mother was very sharp. She had finished the fifth or sixth grade in school—about as far as Greek girls could go in those days—so she could read and write much better than my father. She loved poetry and had memorized many poems, some of them quite lengthy. She also composed long poems of her own, most of them with a touch of humor. She taught me to memorize them, and she also taught me to read Greek. By the age of three I could read the Greek newspaper.

MP: *Amazing. So she thought that reading was important?*

Apostol: There was no question about it. She wanted me to be educated and to do well in school, so she taught me to read, even though there weren't many books in the house. I was a precocious child, so when I was five she thought I should start school. You had to be six to start first grade, and there was no kindergarten in Helper. But she sent me anyway.

MP: *Did you misrepresent your age?*

Apostol: Not very well. When the teacher asked, "How old are you?" I said, "I'm five, but my mother told me to say I was six." They sent me home in tears.

MP: *Was Greek your first language?*

Apostol: Yes. My father was picking up English from his customers at his shoe shop, but he and my mother spoke Greek at home. She began to learn some English from our neighbors. That's also how I learned most of my English, from the kids in the neighborhood and from talking with my first sister. She was born three years after me, and the two of us learned English very quickly.

Math Is Number One

MP: *Did you attend a one-room schoolhouse?*

Apostol: No. Helper Central School was a big two-story brick building with grades one through six. I remember it very well. Looking back, I am impressed at the high quality of the teachers. I had a very good elementary education there.

By the time I reached sixth grade the Depression was going strong, and good teachers had a hard time finding jobs. Many ended up in places like Helper. Our arithmetic teacher was very good and strict; you really had to learn your multiplication tables and arithmetical skills. The principal had the theory that math was number one. If kids were good at mathematics, they would be good at everything else.

MP: *Revolutionary idea!*

Apostol: It was revolutionary. He and the math teacher made up sheets of multiplication and addition problems that we had to solve, day after day after day. It just became second nature after a while.

MP: *Did you like most subjects you encountered?*

Apostol: Yes. I was very good at reading, which made it easy to learn everything. I

Figure 2.3 Apostol (first row, fourth from right) with the sixth grade class of Helper Central School.

also enjoyed learning practical skills as a Boy Scout. I joined at age twelve and became an Eagle Scout at age sixteen. Also, the kids in school were an interesting group to be with. There were Italians, Poles, Greeks, Japanese, and Mormons. Most of their fathers worked either for the Denver & Rio Grande Railroad or in the nearby coal mines.

MP: *Do you think that your mathematical genes come from your mother's side or your father's?*

Apostol: My mother had an uncle who was a mathematics professor in Greece. She herself was very good at arithmetic. My parents often played cards at home with friends, and my mother kept score, doing all the addition in her head. She figured out

Figure 2.4 Tom as a Boy Scout, 1935.

shortcuts by herself, like rounding off to ten and then deciding how much to add or subtract. She had innate intelligence and also good common sense.

She had a keen sense of humor, too, and was a marvelous story teller, with an excellent memory. Sometimes she would retell a story we had heard years before, but I swear she would relate it word for word, exactly as she told it the first time. We always wondered what she would have done had she been more fully educated. She was a remarkable woman and would have accomplished great things.

MP: *The Depression was a hard time to raise a family.*

Apostol: My mother took in laundry to help out. For example, she did laundry in exchange for piano lessons for me and my sister. All four children provided delivery service, picking up and delivering laundry. My mother seemed to be always washing and ironing, or cooking and baking bread. She was a very busy woman.

MP: *And your father?*

Apostol: He worked hard in his shoe shop. One evening a week he would drive to nearby coal-mining camps to pick up and deliver shoes. He loved Helper and the American way of life. Business picked up around 1939 when the Depression began to lessen, and he hired a worker, a deaf man, with whom he communicated by sign language. He also apprenticed several young brothers, the Arlottis, who later started their own chain of shops in Southern California.

Figure 2.5 Apostol with his father and younger sister at the Helper Shoe Shop, 1936.

MP: *So he was a good teacher?*

Apostol: Not really. He tried to teach me to drive, and it was an awful experience. He had tried to teach my mother to drive, and she gave up. He was impatient. If you didn't catch on right away, he would say, "Here, I'll show you. I'll do it for you"—hardly the best teaching method. But his heart was in the right place, and he was certainly hard working and set a good example. My father never took a day's vacation, except to close his shop on Sundays. He was a goodhearted man who continued to send money back to his family in Greece.

MP: *Working six days a week, probably from eight o'clock in the morning to six o'clock at night, does not leave much time for recreation.*

Apostol: We did things together as a family and with other Greek families in the community. We would often take Sunday drives in the countryside, usually up Price Canyon beyond Castle Gate, where we would picnic near springs with fresh running water. That was fun. My mother sang songs while we drove, and everyone joined in. She always livened things up with her songs and her anecdotes. My dad made his own wine in our cellar. He ordered grapes from California, and we wore rubber boots and smashed the grapes in large tubs. I don't know if he learned how from other Greeks or perhaps from our Italian neighbor, who happened to be the town marshal.

MP: *That may have posed a problem.*

Apostol: My dad also had a slot machine in the shoe shop. Every fifth nickel would fall into a drawer at the back of the machine; the rest were used to pay off the players. At the end of the day, my father would unlock the drawer and find at least a dollar's worth of nickels. Sometimes the slot machine earned more than the shoe shop. But slot machines were illegal in Utah, so once a year he would get hauled in and pay the $25 fine. The marshal never confiscated the machines because he knew that people needed the income.

MP: *Your dad sounds like an enterprising guy.*

Apostol: He was. Besides repairing shoes, he sold new ones, and he also cleaned and blocked hats in the shop. He set up all the equipment himself. The naphtha used for cleaning hats was flammable and rather dangerous, but there were no accidents. Eventually he rented a space at the rear of the store for a beauty parlor. His own rent was $55 a month. Some days the shop took in only a couple of dollars, so there were some hard times.

MP: *It sounds like your mother brought a lot of joy into the house.*

Apostol: She was the best thing that ever happened to my dad. We never knew till we were grown up how unhappy she had been when she first came to the United States. My parents never argued, at least not in front of the children. Here were two complete strangers who had a very long and happy marriage. The whole family celebrated their sixtieth wedding anniversary a few years ago.

MP: *It sounds like your parents were a good match. You can look at your mother and see* *some qualities you may have gotten from her. What do you think you got from your dad?*

Apostol: His work ethic, and to some extent I guess you might consider me entrepreneurial.

MP: *Considering all your mathematical projects, I certainly do. As a boy at school, were you interested in math?*

Apostol: Just as I finished sixth grade, Helper built a new junior high school. I had a very good mathematics teacher there, Mr. Pisa, who taught me algebra in the eighth grade and geometry in the ninth.

MP: *Did you take any mathematics in seventh grade?*

Apostol: Yes, but it was a waste of time—more of the arithmetic we had learned in elementary school, with applications to business. Geometry in ninth grade was the most interesting. Mr. Pisa had kids go to the board to present proofs and work geometry problems. He was a hard taskmaster, really severe, but I just loved all of the theorems and proofs. For the first time, mathematics began to make sense, because he followed the classical Euclidean geometry. We started out with axioms and definitions, and it all seemed perfectly logical. And we did the standard two-column proofs. Mr. Pisa could explain things very well, and I always felt he got me off to a good start. For tenth grade I went to Carbon Senior High School in Price, which then became a combination high school and junior college that I attended for one more year. I finished my last year of

high school in Salt Lake City after my family moved there.

Entrepreneurs in Salt Lake City

MP: *What prompted the move to Salt Lake City?*

Apostol: My mom wanted me to go to a university. She saw what happened to the kids who stayed in Helper. They worked either in the mines or on the railroad; that's about all that was open to them. If we moved to Salt Lake I could attend the University of Utah while living at home. We used to visit distant relatives in Salt Lake from time to time. It is a beautiful city, and we liked it a lot. Of course, my dad did not want to move; he was quite content in Helper. He had convenient barter arrangements; for example, in exchange for dental work the dentist and his family would have their shoes repaired. My dad knew almost everyone in town, and he certainly knew all his customers. If we went to the big city, all those personal contacts would be lost. But my mother insisted, and finally my parents took a trip to Salt Lake. They found a combination shoe shop and dry cleaning shop that was for sale. My mother was delighted. She said she could run the dry cleaning business.

MP: *Did she know anything about dry cleaning?*

Apostol: No, but she was smart and was confident that she could learn. After all, she had been doing laundry all those years in Helper. So we moved to Salt Lake, and I helped out in the business, just as I had been doing in Helper, where almost every day after school I stayed at the shop while my father walked home for a coffee break. It wasn't that busy, so I could do homework until a customer came in. I would shine shoes and do minor shoe repairs. I never learned to attach soles on the heavy-duty stitching machine, but I could put on heels. My brother helped, too, when he was old enough. We both helped out in Salt Lake during my last year at high school.

MP: *Was the high school a big change from Price?*

Apostol: It was a very big change. South High was a huge school with 700 in the senior class. I took what they called college algebra, and the teacher was awful. At the beginning of each class she would work the illustrative examples in the text and then assign homework while she sat there and read novels. We spent most of the class time doing homework. When people would ask her how to do a problem, she would send them to me. Since I had studied the book and figured everything out at home by myself, I was able to do all the problems and explain them to my fellow students. The teacher realized that and exploited my talents. She was really bad news, but I enjoyed the subject matter—and I was also happy to be admired by my classmates.

MP: *Did you take any other math there?*

Apostol: The next semester I took trigonometry and solid geometry. The teacher for both was an old gentleman named Mr. Bird who made things interesting and taught us very well. I also had a good chemistry teacher, Mr. Decker. I had taken chemistry in Price, but it was taught by the football coach, who didn't know beans about chemistry. He just had us work problems out of a workbook; then he would read the answers aloud while the students graded each other's work. But Mr. Decker gave interesting lectures and demonstrations. He was really a good teacher. I loved the subject so much that I wanted to become a chemist. So I did very well that year.

MP: *But you had liked math, too, if not as much.*

Apostol: I liked math, and I was always good at it. But I didn't realize at the time that you could make a career out of it. Besides, I was interested in all kinds of things. One was aeronautical engineering. I loved model airplanes, and I built lots of them and sold them to other kids in Helper. When I was about fifteen I sold model airplane kits through my dad's shop. My wholesaler in Salt Lake gave me a 40 percent discount, and to save shipping costs I would hitchhike 100 miles to Salt Lake and back to get the materials.

MP: *So you were a real entrepreneur. Imagine if you had gone to business school!*

Apostol: Unfortunately, our new shop in Salt Lake was just a block away from the model airplane dealer, so I got out of the airplane business. Besides, I was getting ready to go to the university.

Figure 2.6 Apostol with his first gasoline-powered model airplane, 1938.

MP: *So you enrolled at the University of Utah. Was the war on yet?*

Apostol: The Second World War started when I was a sophomore at the University of Utah. There was a military base, Fort Douglas, in Salt Lake, and many soldiers would stop at our shop to have their shoes shined or their pants pressed. This was a gold mine for my parents, because they didn't have to invest in any materials. I helped out with that when I wasn't in class. The university was only fifteen minutes by bus from the shop.

MP: *What was the University of Utah like in those days?*

Apostol: I attended for two years. My best mathematics teacher was Anna Henriques, who taught me college algebra and analytic geometry. She's in her nineties now and lives in a retirement complex in Virginia. I telephoned her recently, and she remembers me very well. I also sat in on a course on complex variables my freshman year. I don't know why; the subject seemed fascinating, but very little of it made any sense to me at the time. In my sophomore year my first semester of calculus was taught by the head of the department, a nice old gentleman who really didn't understand the subject very well. Every time I asked him to explain something he would say, "Work some more problems in the book." I just wasn't satisfied with that. I felt there was something fundamental behind all this formal manipulation, and I wanted to know what it was. The second semester I had a new young faculty member, Frederick Biesele, who was very good. He understood the subject and knew how to explain it.

MP: *So you were interested in math. Was that your major?*

Apostol: I was more interested in chemistry and became a chemistry major. A young chemistry teacher named Hammond, who had just gotten his PhD from the University of Washington, impressed me with his knowledge and teaching ability. I thought, "I'd like to go where he learned chemistry," and I talked to him about that possibility. In 1941, after Pearl Harbor, when everybody began leaving school to enlist, I thought I should join up, too. Since I was still interested in aeronautical engineering, I applied for officers' candidate school in the Air Force. But at the interview they said, "Sorry, we can't take you. Insufficient calculus." Well, I was just beginning to take calculus in my sophomore year, and I was terribly upset. What could I do? I didn't want to join the infantry. Then some administrator at the university said, "Look, the country needs engineers to help the war effort and to rebuild after the war. We'll try to get you guys deferments to stay in school and study engineering." Since I was not too keen on the engineering school at Utah, I switched to the University of Washington, where I majored in chemical engineering.

MP: But your mother's original motivation for moving to Salt Lake was that you could go to the university there and also save money by staying at home.

Apostol: Right. I had to convince her that Washington was a better school for me. I also realized that I was going to have to find work. During my first year at Washington, I was a houseboy in a sorority, with free room and board in exchange for washing dishes for ninety girls; it was great. My dad paid my tuition, which was not very much in those days. In my senior year, I worked nights at Boeing Aircraft. They were mass-producing B-17s, Flying Fortresses, at that time, and every time a wing was assembled someone had to check the alignment of the jigs before the next wing was started. That was my job: an inspector for wing jigs. It was simple, trivial, and periodic. I just waited till someone needed an inspector, then I did my duty. It was great, because I had a desk and could get all kinds of homework done on the job.

Mathematics, by Golly

MP: At the University of Washington, did you study mathematics or only chemical engineering?

Apostol: My BS degree was in chemical engineering, but I had taken so many math courses that I was effectively a math major as well. In my junior year I encountered Herbert Zuckerman, and he turned my whole life around. He taught advanced calculus, and it was the first time I had ever seen advanced mathematics done properly. He explained things well, made it interesting, and made everything look so easy. I never thought of switching to mathematics, but I took number theory from Zuckerman the next year. He used Uspensky and Heaslett as a textbook. There were other good people in the department as well. I took differential equations from C. M. Cramlet, complex variables from L. H. McFarlae, and projective geometry from Roy Winger.

MP: All this while working nights?

Apostol: Yes. When I got my bachelor's degree in chemical engineering in May 1944, the war was still on, and there were no job openings for chemical engineers in the Seattle area. That summer I worked at Kaiser Shipyards in Portland, Oregon, designing layouts for plumbing systems for thirty Liberty Ships. They were mass-produced, but so many modifications were made that no two were exactly alike. As summer came to a close, I decided to go back to the University of Washington as a graduate student in mathematics.

MP: *What motivated you to switch? You still liked chemistry, didn't you?*

Apostol: One day while riding the bus to work, I pondered about my future. I thought, "What am I going to do with my life?" I asked myself, "Of everything I've studied, what is it I like the most?" And I decided it was mathematics. I told myself, "By golly, I'm going to get a master's degree in mathematics and study with Zuckerman!" I thought I would have only a few months of classes before being drafted. But it turned out that my draft board in Salt Lake had misfiled my papers, and I wasn't called for a physical exam until January 1945. By then the Army knew the war was almost over, so they were raising the standards. Since my eyes were bad, I was classified 4-F. And that was that.

MP: *Back to school for you, then. Tell me more about what math was like at the University of Washington.*

Apostol: Most of the male students were military personnel in the V-12 program. There was only one other mathematics graduate student at that time, a fellow named Marvin Stippes. Zuckerman was not eager to teach a course with only two students, so I said, "I'll make a deal with you. I've come across Knopp's book, *Theory and Applications of Infinite Series*, and I'd love to go through it. Why don't we do this as a kind of seminar? Stippes and I will do all the lecturing and you just sit there and keep us on track." Zuckerman loved that.

MP: *Did it actually work?*

Apostol: It was the best training I've ever had. I was the first speaker on day one, and Zuckerman demolished me. He kept asking, "What is the reason for this?"—and I didn't know. "Why is this true?"—and I didn't know. I had read the book carefully and thought I understood it, but it was clear that I was completely unprepared when it came to details. There were proofs about convergence, with epsilons, deltas, and all that stuff that I had never really understood before. That was the last time that happened. From then on, I really studied the book line by line, making sure I understood all the details. Zuckerman was extremely helpful, and I could visit him in his office for assistance. I felt as if I had a personal tutor for a year, learning basic analysis. Knopp's book is marvelous; it has all kinds of great mathematics in it. One of us would lecture each day, sometimes half an hour each, sometimes for the full hour. Zuckerman quizzed us as if we were taking a PhD exam. Nothing got by him. He made sure we understood everything that was going on. Later, when I was a graduate student at Berkeley, my father lost his lease in Salt Lake and bought a new shop in Oakland. My parents loved California; it reminded them of Greece. Anyway, I was working in the shop one day when a customer brought in his shoes, and I asked, "What's the name?" He said, "Knopp." I said, "Oh, are you by any chance related to Konrad Knopp, the mathematician?" "Yes," he replied, "that's my uncle. How do you know about him?" He was stunned. Here was a young guy working in a shoe repair

Figure 2.7 Konrad Knopp.

shop who knew the work of his uncle in Germany!

MP: *What else did you do with Zuckerman?*

Apostol: Every math major there had to write a master's thesis, and I did mine with Zuckerman on magic squares. It was an extension of work done in 1929 by D. N. Lehmer, the father of D. H. Lehmer. It was the first research I had ever done. Years later Zuckerman and I published a joint paper on the topic. It was so exciting to discover something new in mathematics by digging away at it. The first time you discover a new result in mathematics, it is a religious experience. I don't know how else to describe it. I decided that mathematics was the field I wanted to work in for the rest of my life, and it has been very satisfying to do that.

No Books at Berkeley

MP: *After your thesis, were you excited about number theory in particular? Or did you just generally want to be in mathematics?*

Apostol: At that point, I wasn't sure. When I asked Zuckerman about working with him for a PhD, he said, "It's not good to stay at the same school all the time. Why don't you go to Berkeley? D. H. Lehmer is there, and you can tell him about your work with his father's paper." As it turned out, Berkeley had a special one-year scholarship for students from Utah, which I got, and the next year I became a teaching assistant. I showed

Figure 2.8 D. H. Lehmer was Apostol's thesis advisor.

Lehmer my results on magic squares, and he was very pleased. It almost brought tears to his eyes to see the congruences his father had worked on many years before. He said, "I'll be happy to guide your research when you find a suitable problem to work on."

MP: *He left that up to you?*

Apostol: Right. I sat in on his number theory course and registered for his number theory seminar, where I got the idea for my dissertation. Lehmer and I would meet from time to time as I worked on it, but he would just sit there and nod, "Yeah, that looks good. Go ahead, keep working." He did not make a single suggestion about what to do or which way to go. Months later I asked him, "Do I have enough for a thesis? I have to know when to stop." Lehmer said, "Yeah, I think that's enough. Write it up."

MP: *Your experience with Zuckerman probably helped a lot.*

Apostol: The other thing that helped was that they did not use many textbooks in the graduate courses at Berkeley. A lot of good mathematics was being taught, but there weren't any textbooks on analysis or algebra, just lecture notes. The professors were fast lecturers, so we all took notes like mad. I would rewrite my notes at home in little notebooks as a record for myself and as something to review. I would fill in all the details that were not quite finished in class to satisfy myself that I understood the material. This really helped later in writing textbooks, because I learned how to be an expositor, writing so that students could understand.

MP: *Virtually no texts were being used in the graduate courses? How times change!*

Apostol: I do not remember buying a single text. I took a course in complex analysis from Frantisek Wolf, who had just acquired a copy of Widder's book on the Laplace transform. Wolf loved it and began lecturing out of it. After a while, the students formed a little committee and went to see him. "Professor Wolf, this is very interesting, but we signed up for a course in complex analysis." He replied, "This is complex analysis." "Well, it's interesting, but it's not what was advertised in the catalog." I said, "We want a course in complex function theory." "Okay," he said. "You guys prepare the talks, and we'll do it that way."

MP: *The Zuckerman method.*

Apostol: Same thing again. I started preparing some lectures on elliptic functions because this material related to my area of research. And that was when I learned a lot of topics that eventually went into the second volume of my number theory book.

MP: *This explains a lot about your writing.*

Apostol: Very little was available in English in those days. For complex analysis there was Hurwitz-Courant in German, so I studied that and learned German at the same time, which also enabled me to pass the German language exam. I translated portions of Hurwitz-Courant and recast it to fit the modern approach to mathematics that we had been taught at Berkeley, and I started producing lectures on elliptic functions and other topics in complex analysis. I also translated much of the Pólya and Szegő problem books. That was complex analysis. For real variables, I had a very abstract course from Anthony P. Morse, which I didn't particularly care for. Raphael Robinson's symbolic logic course was great; I still have my notes from that. And Cesari was also there.

MP: *The expert on summability?*

Apostol: No, Lamberto Cesari, not Ernesto Cesàro. Cesari gave a course in classical real analysis. He was a character. To begin each lecture, he would call someone to the board and say, "Please tell your colleagues what we discussed last time." This forced students to review their notes before coming to class. If you could not remember all the details, he would say, "Tell us about the theorem on absolute continuity," or whatever, to remind you. He got the students to break the ice and make the transition to his next lecture. Cesari was such a pleasant, outgoing fellow that no one took offense at being asked to do something.

MP: *These sound like lots of strongly supportive teaching experiences.*

Apostol: Yes. Another class that helped me a lot was Alfred Tarski's algebra course. Again, not a scrap of notes and no text—and he wrote very little on the board. On some days he would say, "Let big A be an algebra," and then a big German "A" stood there all by itself. "Okay, theorem one," and then he would write the number "1" on the board— and that was all you saw. He would recite the theorem and proof orally while we wrote everything down as fast as we could. Then,

"Theorem two." And the number "2" went on the board.

MP: *To lecture that way is actually very hard to do.*

Apostol: It is hard, although not all his lectures were like that. Tarski never referred to lecture notes. He had it all in his head. I'm not sure how much algebra I learned in that class, because he played around a lot with axiomatic systems. One day he said, "You can describe an abelian group with one axiom and one operation, division." He gave that as a homework problem, to show that this does indeed define a group. I worked at least forty hours on it! One day he asked, "Has anyone done this problem?" So I raised my hand and went to the board. It took nearly the whole hour to write out the details. He told me later that he had published a paper on this problem years ago in some obscure Polish journal.

MP: *So it was really a research problem.*

Apostol: Yes. He gave us stuff like that all the time, research quality problems. So he was good in making real demands on us.

MP: *How big was the graduate group when you were at Berkeley?*

Apostol: About 200 graduate students. There were about twenty students in Tarski's class. After I got my PhD in 1948 I stayed on another year as a lecturer at the University of California. Then I received a C.L.E. Moore Instructorship at the Massachusetts Institute of Technology and went to Cambridge in 1949.

Boston and Back

MP: *Was MIT a pretty inspiring place?*

Apostol: It was a very good school, and I enjoyed teaching there for a year. I taught advanced calculus and a course in my own field, analytic number theory. Murray Gell-Mann, who was a physics graduate student at MIT at the time and later became one of Caltech's Nobel Laureates, took my number theory course. I had forgotten this until he reminded me of it some forty years later at Caltech.

MP: *Did you get to know Norbert Wiener?*

Apostol: He was quite a character. You would often see him striding up and down the halls accompanied by military people. But he was a terrible teacher. He offered a course on Fourier analysis, one of his fundamental contributions to mathematics, so I sat in on it with Earl Coddington, a fellow instructor. The first day there were two instructors and one student in the audience; after the first lecture the audience went down to zero, because he was such a poor lecturer. He was unaware of who was in the audience, and he talked about whatever he wanted to. He might spend twenty minutes arguing with himself at the blackboard over what symbol to use. But he was a brilliant man. At the weekly colloquium at Harvard, Wiener would sit in the front row and promptly fall asleep, snoring, no matter who the speaker was. The minute the speaker stopped, Wiener would get up and ask a sensible question. As a lecturer, though, he just floundered.

MP: *Did you know Norman Levinson?*

Apostol: Coddington and I sat in on his course in differential equations. Levinson was famous, and the room was packed to overflowing on the first day. But he gave a terrible lecture, just awful, and at the end he said, "This is probably the clearest lecture you'll hear all year." Next time, only about a third of the audience came back, and he began all over again and gave an excellent lecture. That was his way of getting rid of people who did not have a real interest in mathematics. He was a very good mathematician and an excellent lecturer.

MP: *How did Caltech come to recruit you?*

Apostol: I don't know for sure. In a letter to Lehmer about some mathematical question, I mentioned that I did not care for the climate or the living conditions in the Boston area, that I wanted to return to California, and that I had heard of an opening at Caltech. I asked if he knew anything about it, but he never responded to this question. Soon after that, Robert F. Bacher wrote to offer me an assistant professorship at Caltech. I immediately accepted, even though I could have stayed another year at MIT as a Moore Instructor.

MP: *Berkeley was a high-powered place. How did Caltech feel in comparison?*

Apostol: Many faculty members were retiring, and the department started recruiting young, active mathematicians. When I arrived in 1950 the department was very small, with only a handful of full

Figure 2.9 Apostol with Caltech colleague Olga Taussky-Todd, 1978.

professors—E. T. Bell, H. F. Bohnenblust, Arthur Erdélyi, A. D. Michal, and Morgan Ward. Bell retired a couple of years later. But there was a lot of activity because so many young people were being hired. There were parties and get-togethers; the whole department would turn out for picnics. It was a fun place to be. What I liked best was the high caliber of the students. If anything has kept me at Caltech all these years, it has been the quality of the undergraduate student body.

MP: *What were you interested in doing when you arrived?*

Apostol: I didn't have any grandiose plans. I wanted to be a good researcher and a good teacher, and the atmosphere here was very good for that. I had just published my PhD thesis and was making good progress on

related topics. The teaching load was fairly heavy. We had to teach two courses, which takes a lot of time when you are starting out. I often taught courses I had never taken myself, so I had to learn the material and teach it at the same time. I also used to participate in seminars, both actively and passively. E. T. Bell and I conducted a private seminar for just the two of us for a couple of years before he retired.

Calculus for Caltech

MP: *What made you decide to write your first calculus text?*

Apostol: I was asked to teach the advanced calculus course, which had been using volume two of Courant's *Differential and*

Integral Calculus from the 1920s. It was a very good book, but too low-level for that kind of course. There was no book in English that was intermediate between elementary calculus and real variable theory, sort of an introduction to both real and complex analysis. When Morgan Ward and I were assigned to teach the course for 1953 and 1954, we couldn't find an appropriate text, so I said, "Morgan, you and I could probably write such a book." We made a tentative plan to write up a few chapters over the summer and then get together in September to see what we had come up with. I spent the summer in Oregon working like crazy, and when I came back I found that Morgan had forgotten all about it. So I offered to write up a set of lecture notes by myself, which I did. I managed to keep three months ahead of the students during the year. Warren Blaisdell, who was then vice-president of Addison-Wesley, heard about those notes and asked if he could have them refereed, with the idea of publishing them as a book. I said, "Well, you know, they're just sketchy notes. They're in no form for a book." But he got some very enthusiastic responses from the reviewers and sent me their critiques. So I spent another summer working like crazy to transform them into a publishable manuscript. *Mathematical Analysis* was published in 1957 and became somewhat of a classic in its field. A whole generation of mathematicians was raised on that book, and it has been translated into several languages.

MP: *Tell us about the process you went through in writing your two-volume* Calculus *textbook. You had weekly development*

Figure 2.10 Apostol after his promotion to full professor, 1962.

meetings with math colleagues and physicists, I understand.

Apostol: It took about five years overall. Caltech students were coming here better and better prepared—not only bright but with a lot of background. There was no book on the market that was suitable for these more sophisticated kids. First we tried a preliminary edition of George Thomas's *Calculus*, which used vectors and other things that had never been in a calculus book before. But the students didn't like it, so we decided to write our own book. Some of us sat down and discussed what would make sense to teach these bright kids. We considered all kinds of possibilities, but in the end we decided that calculus was the best thing we could offer them. We met for a year, one hour every week, thrashing ideas around until we developed a sort of master plan for an integrated course in freshman mathematics. The physicists were made aware of the changes, why we were doing things differently, and how the mathematics would mesh with freshman physics. Then came the question of who would transform our plan into useful notes. I had already published one book, so I had some experience. And actually I was eager to do it! So I wrote little booklets of mimeographed notes, keeping one term ahead of the course, just following the plan from our meetings, filling in all the details, and adding suitable exercises. At the end of the year it dawned on us that we had

Figure 2.11 Apostol delivering a transparent lecture.

not discussed the follow-up course in sopho-more mathematics. So I modified the Math 2 course to fit what we had done in Math 1, and again it was discussed with the physi-cists. Then I wrote another set of notes the same way as before, keeping one term ahead of the students. This eventually became my *Calculus*, volumes one and two.

MP: *It must have been well received by the students, since they refer to your calculus volumes as "Tommy 1" and "Tommy 2."*

Apostol: They knew this was something new, and they liked the approach. There was a sense of excitement, because this was "the book written for Caltech." It took into account the needs of other departments,

physics primarily, and it also satisfied the mathematicians, who were happy to teach this material.

MP: *How important is it for mathematics students to learn physics?*

Apostol: To have some kind of physical feeling for what things mean—that's good for everybody. I have always felt that applica-tions are an important part of a mathemat-ics education. Don't forget, I'm an upward bound engineer! And I had a minor in physics when I got my MS at Washington. That's why my calculus book has a lot of applications to physics and engineering. Anything that helps increase understand-ing and insight is important. If you can see

Figure 2.12 Tom and his calculus class celebrating Gauss's 133rd birthday, 1988.

Figure 2.13 Tom and Jane Apostol hamming it up at a Democratic fundraising party in 1960.

something from two or three points of view, that is valuable. I try to give my freshmen the message that great achievements are made by people who see these connections, not by those who specialize in just one field.

MP: *So they should be willing to study even abstract concepts, because they might apply someday?*

Apostol: Yes, although it is difficult to get freshmen to feel that way. They want something utilitarian, and they don't really care why it works, as long as it works—like driving a car. It is a constant challenge to convince them that you also have to think about why it works. What if your car breaks down?

MP: *Other departments that require mathematics often complain, "The math department is just completely out of touch with us." Many institutions have people teaching calculus who probably never had anything beyond a high school physics course.*

Apostol: It is important for mathematics students to understand how mathematics is used in other fields. You are depriving them of an education if you don't teach them that.

MP: *Your teaching experience at Caltech has largely been positive, in part because the students are good. What makes a good teacher?*

Apostol: Output is proportional to input, so if you work hard at it, you can do a good job. If you are not interested and don't prepare your material, you won't do a good job—unless you are an exceptional person. But I always put a lot of effort into the courses I taught because I felt it was important. Mostly, I wanted to make sure that what I said was understandable to the students and that it motivated them to want to learn more. I tried to anticipate their difficulties. Remember, sometimes my teachers in high school and college did not understand what was going on, and it disturbed me when they could not explain it. I was always interested enough in mathematics to try to understand everything, and in my books I always tried to communicate my own concerns to the students.

The Grecian Formula

MP: *Have you ever visited your parents' homeland, Greece?*

Apostol: Many times. My wife and I first went there in 1962 as tourists and then again in 1965. I met some Greek mathematicians on these visits and was invited to be a visiting professor at the University of Athens in 1967. But the military dictatorship took over, and since the Greek government runs the universities, I did not want to work for someone who suppressed academic freedom.

MP: *Did the best Greek mathematicians tend to stay in Athens?*

Apostol: The young talented people would usually leave Greece. The universities followed the old German system that everyone else in Europe gave up long ago. The professors were autocratic and had a lot of power. People with talent couldn't break into the system or get promotions unless they had family connections or some tie with the government. The situation improved when new legislation was passed in the 1980s. In 1978, after the dictatorship ended, I went for a four-month stay at the University of Patras, a new school. I worked hard preparing to deliver my lectures in Greek—and when I arrived the students were on strike, so there were no classes.

MP: *Why were they on strike?*

Apostol: It seems like a strange reason. They wanted the right to be promoted from year to year even if they had not passed any exams. Under their system, once you were admitted to the university, you could stay forever. Students got a daily food allowance, free books, free tuition. Most had outside jobs for extra income and just went to class occasionally. But their outside salary depended on the class they were in, so they all wanted to be upperclassmen to get more money.

Figure 2.14 In 2001 Apostol was elected a Corresponding Member of the Academy of Athens. In May 2001, he gave his inaugural lecture to the Academy.

I ended up just giving a small seminar in English every week to some faculty members. This was not announced publicly for fear that the striking students would disrupt it. The only undergraduate I talked to was a young lady who was thinking of going to America and wanted my advice.

MP: *Was there any movement to change the system?*

Apostol: In the early 1980s new legislation was passed to establish a new university in Crete. It was going to be set up more like the American system: departments instead of chairs for life, department members voting on new members, and so on. I served on an electoral committee that selected the initial mathematics faculty, based on merit. We met eight to ten hours a day in grueling discussions to fill all the positions established by law. When excavations began on the site for the new campus, antiquities were unearthed, and construction was halted. Of course, there are antiquities everywhere in Greece. So this grand new place had to start out in temporary buildings. The mathematics office was once located in downtown Irakleion directly above a pizza parlor. In time, the academic atmosphere improved throughout Greece as new legislation was passed requiring all Greek universities to emulate the Cretan model. Some new universities have been formed in outlying provinces, but political considerations seem to play a role in the way they are structured because everyone wants to share in the benefits that flow from government funding. For example, a new campus may be fragmented so that the engineering school is in one province but the medical school is in another. It is difficult for the faculty to exercise control over a university system that is owned and operated by the Ministry of Education and Religion. But the Greeks are intelligent and hardworking, and I'm sure they will figure out a way to overcome these difficulties.

Project MATHEMATICS!

MP: *Ten years ago you started the award-winning video series, Project MATHEMATICS! How did you get involved with film making?*

Apostol: It all began in 1982 when a Caltech physicist, David Goodstein, invited me to be part of the academic team for The Mechanical Universe, a fifty-two-episode telecourse in college physics that was calculus based. Goodstein wanted me aboard to make sure the physicists didn't botch the mathematics, but as time went on I became more heavily involved, both as a scriptwriter and as a co-author of three textbooks. The turning point for me was when I saw the first example of Jim Blinn's computer animation. It was clear that this was a new and powerful tool that could be exploited for mathematics education. It brought mathematics to life in a way that cannot be done in a textbook or at the chalkboard. I was hooked, and for the next five years I devoted nearly all my energies to this physics project, learning new skills such as scriptwriting, television production and editing, the use of original historical documents, and other techniques well known to the television industry but rarely encountered among research mathematicians.

As the Mechanical Universe project began to wind down in 1985, I asked Blinn if he would be interested in working on another project for mathematics at the high school level in which his animation would play an even greater role. He agreed, and the rest is history. Suddenly I had to learn a new skill, fundraising. It took a long time and a great deal of effort to find money to launch Project MATHEMATICS! The success of The Mechanical Universe helped me to obtain some seed money from the Caltech administration to get organized. I wrote an article for *FOCUS* in 1986 describing our plans, and this helped generate our first major grant from SIGGRAPH. The rest of the story is described in an update I wrote for *FOCUS* in 1994. By the way, there is an amusing story about the name of the project. In the 1986 *FOCUS* article it was called Project MATHEMATICA, and our first video was produced under that name in May 1988. In June 1988 Steve Wolfram announced his new software called MATHEMATICA. I immediately telephoned to let him know we had just completed a video under that same name and asked if he could choose another name, because some people might confuse the two projects. He declined, explaining that he had already spent a small fortune registering a trademark for MATHEMATICA in several countries. To avoid a legal dispute, I notified Wolfram a few days later that we were changing our name to Project MATHEMATICS! (with an exclamation point). He replied, "I wish I had thought of that. I never did like the name MATHEMATICA."

MP: *What have been the joys of making videos? The low points?*

Apostol: Of all the things I've done in my fifty years as a mathematician, the most satisfying has been producing and directing the videos for Project MATHEMATICS! All mathematicians love to solve puzzles. Making a video is like solving a puzzle. You have all the ingredients for a great story—the mathematics itself, heroes and history, applications to real life, computer animation, and a vast resource of visual images. The challenge is to blend all these together with a sprinkling of music and sound effects to reveal the beauty of mathematics and to provide understanding and motivation for learning more. The enthusiastic response we have received from viewers of all ages and from all walks of life tells me that we have succeeded in creating something of lasting value.

The down side of this endeavor is the constant search for funds to support the effort. Fundraising takes a great deal of my time and energy that could be better spent in more creative ways. The funding problem was the primary reason that Jim Blinn left

Project MATHEMATICS! and went to work for Microsoft.

MP: *So, over the years, you have been a shoe repairman, a houseboy, a teacher, a writer, a curriculum developer, a maker of videos, and in Greece an administrator of sorts. Like your parents, you keep busy! What do you want to do next?*

Apostol: If I had my way, I would continue to make mathematics videos. But it's not easy without an animator like Jim Blinn to work with. Currently we have a grant from the National Science Foundation to produce an interactive version of Project MATHEMATICS! The idea is to provide discovery-based interactivity where students can discover things like the Pythagorean theorem by themselves. This is a completely new challenge, both pedagogically and technologically. Fortunately, I have a couple of creative people working with me, Benedict Freedman and Mamikon Mnatsakanian, and together we will give it our best shot.

MP: *Tell us a little about your mathematical research.*

Apostol: Most of it has been in analytic number theory. Number theory deals with the integers—prime numbers and other special sequences of integers. Analytic number theorists study them using tools from real and complex analysis, such as Dirichlet series, infinite products, and elliptic functions. Most of my work has been on these tools and their interrelationships. It is the kind of research that is competent and needed, but if I didn't do it, someone else would.

Looking back on my career, I think my strength lies in mathematical exposition rather than research. I have tried to tell students what is going on through my books and videos in a way that is both interesting and understandable. I've always followed the old principle: Tell them what you want to do, do it, and then tell them what you've done. It's a very good formula to use in teaching.

Three

Harold M. Bacon

Gerald L. Alexanderson

Many great research universities have on their mathematics faculties one or two senior people who do not spend their lives producing important theorems but still contribute greatly to mathematics, both on their own campuses and more widely. They are often revered by their students for their classroom teaching and their mentoring. One such was Harold Maile Bacon, who, with his longtime colleague Mary Sunseri, played just such a role in the mathematics department at Stanford University.

Bacon came to Stanford as a freshman in 1924, following in the footsteps of his father who had graduated only a few years after Herbert Hoover, when Stanford was a fledgling institution at the end of the nineteenth century. Harold stayed on for graduate study, receiving his PhD from Stanford in 1933 under the direction of the eminent Danish analyst Harald Bohr, who was visiting Stanford at the time. Bohr, however distinguished as a mathematician, may have been better known as a champion soccer player in Denmark and as the brother of the Nobel laureate physicist Niels Bohr. Bacon, after a short stay at San Jose State, returned to Stanford in 1935 and rose

Figure 3.1 Harold Bacon was a legendary mathematics teacher who taught at Stanford for forty-five years.

through the ranks to promotion to full professor in 1950. He formally retired in 1972 but continued to teach until 1980.

A Legendary Teacher and Advisor

At Stanford he was a legend, having taught calculus and other subjects to generations of

Figure 3.2 George Pólya and Harald Bohr, brother of Neils Bohr and Bacon's thesis advisor.

Figure 3.3 The young Bacon family—Harold, Rosamond, and baby Charles.

Stanford students. In 1965 he was given Stanford's Dinkelspiel award for distinguished service to undergraduate education. Everyone I've ever talked with about mathematics at Stanford during Bacon's years there remembered him, if not for having taken one of his classes then for having received from him the best advice on majors, careers, and life in general.

He was a faculty member of the old school, formal in demeanor, always in suit and tie, often with a vest. He was somewhat courtly, polite, gracious, welcoming, and friendly. Yet he was always reserved, well aware of the distinction between faculty

members and students. He married, rather late in life, another person deeply committed to Stanford, Rosamond Clarke, and they had one son, Charles, now a prominent geologist. They lived in the Dunn-Bacon house on fraternity row, near the famous Stanford Quad. It is a large neoclassical house that once belonged to Harriet Dunn, a cousin of his father's. The interior was paneled and formal, with Victorian lamps and furniture consonant with Stanford's Victorian origins. On the third floor was an elaborate model train layout for Charles. In front of the Ionic columns around the front door, a curved driveway was lined with rose bushes, where the Bacons were often seen tending their roses, Harold usually in attire more formal than that typically seen on gardeners.

Harold and Rosamond were the quintessential faculty couple, involved at all levels of campus life, known to everyone in the Stanford community, generations of alumni, and members of the Palo Alto community at large. In some ways they were the

Figure 3.4 Charles Bacon, son of Harold, went on to a career in geology.

models of what one might at that time have thought of as the perfect Republicans—Episcopalian, socially well connected, and, in their lifestyle, conservative, indeed almost patrician. Of course, as often happened in their day, even with that lifestyle they were also politically liberal. Harold once told me the secret of getting through life is to trust in God and vote the Democratic ticket.

Rosamond was throughout Harold's career no passive partner. She had worked for the university after she graduated and was an active member of the Library Associates and a founder of the Stanford Historical Society. John E. Wetzel, now professor emeritus at the University of Illinois, Champaign-Urbana, recalls that when he came to Stanford "the Bacons hosted a reception for new graduate students shortly after [he] arrived . . . and [he] remember[s] a wonderfully pleasant chat with Mrs. Bacon as she served tea . . . [S]he was so interested, friendly, and welcoming that I felt a real sense of loss when she left me to turn to someone else."

Teaching Calculus to an Alcatraz Prisoner

Always involved in doing good works, the Bacons often helped unfortunate people in the community, even on occasion to the point of finding lawyers to get them out of jail when they were unfairly detained because of their background. In a much more serious case, well known to the campus community, beginning in 1950 Harold made trips to Alcatraz by launch from a pier in San Francisco to help a prisoner who was trying to learn calculus on his own. Alcatraz—"The Rock"—had housed some of the most hardened criminals in the United States, including Al Capone and "Machine Gun" Kelly. The prisoner, Rudolph "Dutch" Brandt (Reg. No. 369-AZ), had been convicted of killing a gambler in 1924 and subsequently of participating in a bank robbery in Detroit in 1936, after which he was sentenced to a twenty-five- to thirty-year term.

Brandt was almost completely self-taught, but, because there were no provisions for taking courses at Alcatraz, he enrolled in a correspondence course in calculus from the University of California. He had found working on algebra and trigonometry problems beneficial while in solitary confinement, a period that Bacon estimated to be between one and three years. While in what Brandt called the "dungeon," he had essentially nothing to write on so he tore up pieces of tissue paper to form letters and numbers so he could work out a method of finding the solution of four simultaneous linear equations. At one point Brandt wrote apologetically to Harold indicating that he was sorry to take up so much of his valuable time. Harold wrote back: "Do not feel that [my] coming to see you is in any way a waste of my time regardless of what use the mathematics may be to you when you get out of the penitentiary. The greatest value to be found in your study of mathematics is, it seems to me, the satisfaction you get from the subject itself with the accompanying sense of achievement."

Correspondence and visits continued until Brandt was transferred to a federal prison in Michigan, where he was allowed to develop skills working in the prison shop. Paroled in 1953, Brandt worked in a tool and die shop in Detroit, eventually moving to Cleveland. He continued to correspond with Bacon and on a trip east to attend a meeting of National Science Foundation (NSF) institute directors, Bacon stopped off to visit with Brandt, who was by that time suffering from terminal cancer. Much earlier Bacon had taken him books of tables to help him with his calculations and once gave him a specially autographed copy of George Pólya's *How to Solve It*. For Christmas 1953, Rosamond sent Brandt a Stanford Christmas calendar and a picture of their son Charles, then five years old. Brandt, who earned only a meager salary, sent a $50 bill to Charles for Christmas, ultimately used to buy a new piece for his model train setup. Brandt died in late 1956. Details of this friendship were chronicled in some detail in a moving article by the Bacons' friend Charles Jellison in "The Prisoner and the Professor" (*Stanford Magazine*, March-April 1997, pp. 64–71).

Harold had been alerted to Brandt's efforts to learn calculus by Lester R. Ford, then at the Illinois Institute of Technology and a former president of the Mathematical Association of America. Ford had been corresponding with

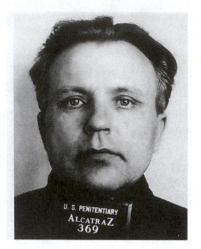

Figure 3.5 Rudolph "Dutch" Brandt, a prisoner at Alcatraz. Bacon made regular trips to the prison to teach Brandt calculus.

Brandt and had encouraged him to look at some mathematical problems. After visiting Brandt, Harold wrote to Ford to report on their encounters in the prison. He said, ". . . he is eternally grateful to you for what you have done for him. It means everything that you have addressed him like a man rather than a convict. He spoke feelingly (and there was a world of genuine pathos in this poor fellow) of his struggles to conquer the mathematical questions he had tackled. He now feels that he is winning his fight to learn. In this he sees his chance to get clear of the Underworld (his own words)." Jellison went on to say that "Bacon treated him as he did all his students, with respect and compassion. He shared his love of mathematics, 'that handmaiden and queen of the sciences,' with a murderer and a bank robber, and was rewarded by witnessing his transformation."

Bacon carried on correspondence with prominent mathematicians (people wrote real letters in those days), notably Dunham Jackson, well known for his American Mathematical Society (AMS) Colloquium Series volume, *The Theory of Approximation* (1930) and his MAA Carus Monograph, *Fourier Series and Orthogonal Polynomials* (1941), as well as Harold Davenport, the eminent number theorist at Cambridge University. Correspondence with Jackson consisted, in part, of an exchange of their own limericks, albeit very proper ones.

Neither Harold nor Rosamond came from a family of academics. Harold's father was an engineer for the city of Los Angeles and designed pedestrian walkways under city streets near public schools. Rosamond's family came from Carpenteria near Santa Barbara. Her siblings married well. Her brother, Thurmond, was at one time the Chief Judge of the U.S. District Court for Los Angeles, and he married Athalie Richardson Irvine,

Figure 3.6 Professor Bacon enjoying lunch by the pool.

the widow of James Irvine, Jr., of a pioneer California family that had developed much of Orange County, specifically the Irvine Ranch that lent its name to a campus of the University of California. Her sister, Sue, married a member of the Roos family, a prominent mercantile clan in San Francisco whose house on Pacific Heights is an architectural icon designed by the famous Arts and Crafts architect, Bernard Maybeck. Sue's house in the fashionable suburb of Belvedere extended over San Francisco Bay and was called "The Stilted Manor." Rosamond married an academic.

Dickens and Thackeray

Aside from their good works and his campus position, Harold and Rosamond lived rather quiet but comfortable lives, reading to one another at bedtime from their favorite works of Dickens and Thackeray, and they listened to Gilbert and Sullivan.

Always traveling either by car or by Pullman across the country, in the 1960s, after taking the train to New York, they sailed on the Queen Mary to England, where they stayed at Brown's Hotel in London. It had opened in 1837 and proudly displayed, even in the 1950s, the rent in the lobby carpet where Elisabeth, Queen of the Belgians, once caught her heel when she lived there during World War II.

I first met Harold when I showed up on the Stanford campus in the summer of 1955 when I decided to take a look at housing and such just before my first year of graduate study there. No one else was around, so I

Figure 3.7 Harold Bacon and Harold Davenport at Stanford University.

wandered into Harold's office (he was usually there whether school was in session or not), and he greeted me warmly and provided me with excellent suggestions on housing and what I might expect in my first attempt at teaching. After that first encounter I always felt that we were friends.

Pioneered Institutes for High School Teachers

In the 1950s and up to the mid-1960s, Harold was a pioneer in introducing, with funding from the National Science Foundation and the Shell Foundation, a series of summer and academic-year institutes for high school teachers. The idea had been developed during conversations on a long drive with George Pólya to attend some meetings in Colorado. For these programs, which he started in 1955, he brought in outstanding faculty—H.S.M. Coxeter from Toronto, Carl B. Allendoerfer of the University of Washington, I. J. Schoenberg of the University of Wisconsin at Madison, Morris Kline from New York University, Ivan Niven from the University of Oregon, and D. H. Lehmer of UC Berkeley. The high point was a series of two NSF institutes held in Versoix, a village just outside Geneva, Switzerland, for teachers in American schools abroad. As a young faculty member I was fortunate to be asked to assist in the administration of these two institutes and to be able to teach in both of them, in 1964 and 1965, where Harold and I were joined one year by George Pólya and the other by Cecil Holmes of Bowdoin College. Our assistant was a recent PhD at the Eidgenössische Technische Hochschule

Figure 3.8 Members of the NSF institute faculty in Versoix, Switzerland in 1964–65—George Pólya, Harold Bacon ("Monsieur le director général"), and Cecil Holmes of Bowdoin College.

(ETH) in Zürich, Pia Pfluger (later Korevaar), who went on to a career at the University of Amsterdam; she was the daughter of Pólya's successor at the ETH, Albert Pfluger, when Pólya left Zürich in 1942. The administrator of the local Collège du Léman, where the institutes were held, referred to Harold as "Monsieur le directeur général" and to me as "Monsieur le directeur adjoint." Happily, the instruction was in English.

When not in the classroom, we had to struggle along with our college French, which got us through from day to day except on a few occasions. When Harold backed our leased Peugeot into a wall and we had to negotiate with local repairmen to get the fender fixed, all we could do was point, since "fender" is not in Berlitz and is seldom heard in a college French course. And Harold, more inclined to go to church on Sunday mornings than I was, felt that part of my task of being "directeur adjoint" was to accompany him to the local Old Catholic Church, a nineteenth-century

Figure 3.9 The "birthday lineup": Albert Pfluger (October 13), Gerald L. Alexanderson (November 13), George Pólya (December 13), and Harold M. Bacon (January 13).

Swiss breakaway from the Roman Catholic Church, when going to the Anglican church in Geneva was inconvenient. He was more successful at appearing to know what was going on during those services than I was. I do recall, though, that one of the favorite hymns in the local church was sung in French to the tune of Stephen Foster's "Old Folks at Home," otherwise known as "Suwannee River."

Of course, Harold's influence was not restricted to his own teaching or his directing these teacher institutes. He wrote a popular calculus text (McGraw-Hill, 1942; second edition, 1955) that was much admired for its clarity and its beautiful problem sets, along with *Introductory College Mathematics* (Harper, 1954), co-authored with Chester Jaeger of Pomona College. Of the calculus text, Wetzel said, "I learned my calculus the first time I taught the course, from the [second] edition of Bacon's *Calculus*, a book I used

as a problem source throughout my teaching career." Harold also published a beautiful piece on Blaise Pascal in *The Mathematics Teacher* (April 1937, pp. 180–85). One of the organizers of the Northern California Section of the Mathematical Association of America (MAA) in 1938, he served as the section's first secretary-treasurer and remained active in the section for many years. He was elected twice to the MAA's national Board of Governors and served on many of its committees, in addition to countless committees on the Stanford campus. An exemplary citizen of the campus and various mathematical communities, he was known widely in professional circles during his career.

In mathematics education he had great enthusiasm for the work of George Pólya and was the first person to call to my attention an extremely short and elegant proof by Pólya of the Cauchy-Schwarz inequality, using only the exponential function and the first two terms of its Maclaurin expansion.

Recently I had the opportunity to talk with a couple of experienced mathematicians who, like me, were teaching assistants at Stanford in the late fifties. We were amazed in comparing notes on our teaching over, now, many years to see how similar our practices have been and how much we learned from Harold Bacon about teaching, even down to the mechanics of grading, setting policies for homework and examinations, advising students on suitable courses, and answering their questions on careers. For example, he advised showing students scores on midterms and finals, listed in rank order, without assigning letter grades to any scores, so that students were not tempted to "average" letter grades, usually to their advantage. And he advised against any preset equivalence of ranges of cumulative totals of points and letter grades, since that presumed that the instructor always made up near perfect sets of examination questions. This goes a long way toward discouraging extended arguments with students about the final course grades. (They like to use the Peppermint Patty system from the comic strip Peanuts: "The way I see it, seven 'D-minuses' average out to an 'A.'") As a department chair

for over thirty-five years in my own department, I have seen many examples of what trouble young instructors could get into by ignoring these simple procedures.

Sometimes Harold's advice was just plain shrewd: when Howard Osborn was a beginning teaching assistant at Stanford, he faced the common problem of having a premed student who claimed after every test that she had to have an A "in every one of her courses because no medical school would accept her otherwise." Harold advised him to agree to reread all of her quizzes and exams "very carefully ... [and] find additional minor errors that were ignored during the earlier grading and that deserved a lower grade." That tended to discourage further arguments. What we learned from Harold in those days as teaching assistants served us well through many, many years of teaching. In turn we have passed along to our own students his wise counsel.

In my experience, Harold Bacon's legacy is very similar to that of many others on research university campuses, and their influence is long felt, even though their work is not recognized in the mathematical encyclopedias. They have been, however, truly outstanding mathematical people.

Four

Tom Banchoff

Donald J. Albers

As an undergraduate, Tom Banchoff missed the class of one of his teachers, a distinguished mathematician, and made the mistake of apologizing in the hallway. His teacher was infuriated by his absence and, in a very loud voice, told young Banchoff, "You will never be a mathematician, never, never!" That teacher was wrong, for Tom Banchoff has gone on to a distinguished career as a researcher and teacher. He is now Professor of Mathematics at Brown University.

As a high school student, Banchoff, after reading a Captain Marvel comic book, became interested in the fourth dimension and *Flatland*, a book by Edwin Abbott Abbott that has had a huge influence on his work. In the early 1970s, he was producing pioneering films that illustrated the fourth dimension. His passion for and involvement with the fourth dimension has continued unabated for more than fifty years. During that time, as technology has evolved, his illustrations of higher dimensions have improved dramatically. Over that same period, his appreciation for Abbott has grown to the point that he and William Lindgren have written the quintessential guide to *Flatland*, published jointly in

July 2010 by the Mathematical Association of America (MAA) and Cambridge University Press. He also has written *Beyond the Third Dimension* (Scientific American Library, 1990) and *Linear Algebra through Geometry* (second edition, Springer Verlag, 1991).

From an early age Professor Banchoff knew that he wanted to be a teacher. Even a brief conversation with him reveals that his dedication to teaching is extraordinary. Throughout his career at Brown he has involved students in research projects, many of which were centered on higher dimensions. He speaks with great pride about his students' accomplishments, which stand them in good stead for graduate school and/ or employment.

Banchoff's commitment to students is reflected in his winning several teaching awards, including the MAA's Award for Distinguished College or University Teaching (1996) and the National Science Foundation Director's Award for Distinguished Teacher Scholar (2004).

Professor Banchoff served as President of the MAA from 1999 to 2000, and he is the recipient of two honorary doctorates.

Banchoff: My father never pushed me to become a scholar, although he impressed upon me the fact that learning English and arithmetic (I didn't hear the word "mathematics" until I was in high school) were the sorts of things that are really important, and none of the other things are. He was also very conscious of the fact that I should be "a regular guy." So I took on the additional responsibility of becoming a regular guy. All these other things that I did, such as reading, he couldn't complain about.

MP: *What did he mean by being "a regular guy"?*

Banchoff: He meant that you shouldn't be eccentric, like a book worm, studying all the time. He was actually rather suspicious of intelluals, and he didn't particularly like them. He knew people who read books and talked about books. He, himself, wasn't a reader. My mother, on the other hand, read all the time, and she was very encouraging. There was a little bit of tension there, I suppose. A middle-aged reclusive neighbor was my father's example. He said, "You don't want to grow up like him." He worked at a library and used to carry his laundry in a paper sack.

MP: *Did you become an all-around guy?*

Banchoff: Yes. I was on the tennis team, and I played soccer. I was in the school play, the school orchestra, the band, and I was on the debate team. I was editor of the school paper and a member of the yearbook staff. I was in the Third Order of St. Francis and president of the Latin Club.

Figure 4.1 Banchoff, on his fifth birthday in red boots, trying out his new toy fishing rod.

How to Get Perfect Grades

MP: *Where did you attend school?*

Banchoff: Trenton Catholic Boys' High School in Trenton, New Jersey. I ran a 99.29 average for four years. Most people saw me totally involved with other activities all the time. I didn't want them to think I was studying or something like that. It was one of these silly situations in which there was a very undemanding curriculum. At the same time, I apparently got the idea that the way you should approach school is to figure out how to get perfect grades in all the examinations and just do that much.

MP: *How did you go about figuring that out?*

Banchoff: That's easy: pay attention.

MP: *To your homework?*

Banchoff: To your homework, yes. That never took any time. I liked it.

Figure 4.2 Banchoff, age 11, in suit, tie, and hat, about to take his first flight. He was off to Chicago to be a contestant on the "Joe Kelly & The Quiz Kids" radio program.

MP: *What subjects did you particularly enjoy in high school?*

Banchoff: Mathematics, of course, and languages. I was interested in the structure of languages rather than literature. There weren't any subjects that I didn't enjoy, although I didn't care much for laboratory science. And I didn't read enough to decide to go into English. I had already decided that, since I hadn't read Thackeray, there was no sense being an English major.

MP: *You were taking Latin.*

Banchoff: And German, and I studied French on my own.

MP: *On your own?*

Banchoff: One of my teachers told me that French was better than German, so when I didn't have a job one summer, I did all the exercises in my teacher's French text.

MP: *That's fairly uncommon. Did you know any other students doing that?*

Banchoff: No, but I knew a couple of the students who were studying French. I used to work with them. I never learned to speak French correctly. It's a total embarrassment. I also was very interested in philosophy and theology. But I had been counseled by people that you don't do that as an undergraduate; you do that later in life.

MP: *Who provided that counseling?*

Banchoff: Well, my teachers. I was taught by Franciscan priests. And they were right after all.

MP: *They should know.*

Banchoff: Yes. They never pushed me to become a priest. And I think I'm really better doing what I'm doing. The idea that I would go off to a seminary was never pushed, which was nice because in a certain sense in those days good students were encouraged to become priests.

MP: *Do any high school teachers stand out in your memory?*

Banchoff: I knew one special teacher, Father Ronald Schultz, when I was a freshman. He believed the best of you, even some of these guys who were questionable in my estimation. He really was a spiritual father, although he never taught me. He was also a math teacher, and he listened to people. He actually listened to me. He was the head of the Mathematics Department, and he was also the moderator of the Mass Servers Society, which was a kind of fraternity. It was an all-around-guys kind of fraternity. He'd get up at five in the morning and pick us up, and we'd go and serve mass for the twenty-eight priests. Then we'd sit around playing cards or doing homework or whatever, waiting for school to start. It was a real fraternity and a really good group of guys. Some of the toughest kids in the school belonged to it. It was really a very interesting experience, and everybody loved Father Ronald, who was the kind of person who would just go out of his way to stick up for you.

The Fourth Dimension

I proved my first geometry theorem when I was a freshman because we used to have to go to mass every Friday morning, and being in homeroom 1-1, we filed in first, so we were way up in the first pew. You'd have to wait until the other 800 kids came in. So you had a lot of time to pray or to contemplate or to watch shadows move across the tiles. There were square tiles set obliquely across the altar rail—we used to have altar rails in those days—which would cast a shadow. By the beginning of the mass, it was very thin on one side, and by the end, it would cover almost an entire triangle so I posed to myself the question: when was the shadow exactly half way across? Not half way in terms of distance, but half way in terms of area. I hadn't had geometry yet, of course, but I figured it out. When it's half way across, you get an isosceles right triangle, which is half of what you get if you cut it the other way, so just cut it in half, and you've got it. I remember once explaining this to one of my friends at an evening meeting of the Knights of the Blessed Sacrament, and he said, "Father Ronald, the new kid here has a theorem." "Well, what is it?" "Here's the problem." He said, "Well, you'd have to set up a relation and a proportion and solve some algebraic equation." I said, "No, I didn't do it that way." He said, "If you already knew the answer, why did you ask me?" I said "But I didn't ask you." And then, fortunately, something happened to distract him, but we both remembered that later on. Anyway, he was the only one who listened to me because I was really very much into the fourth dimension at the time.

MP: *As a high school student?*

Banchoff: Yes. When I was in junior high, I was into it. I was really interested in dimensional analogies and theological mysteries. How you could understand time and how you could understand how a circle could be a straight line, etc. I was into the Flatland analogy, not really through "Flatland" so much but comic book versions of it. I'd like to think Father Ronald was the one who gave me "Flatland." He never told me that I was asking dumb questions. At the end of my freshman year, he was transferred, so I never had him as a teacher. I remember one thing, though. Once my freshman algebra teacher gave us a number of things that we were supposed to memorize. One of them was every number multiplied or divided by zero equals zero. I couldn't wait. I ran up to where the teachers used to smoke between classes. Father Ronald said to me, "Tommy, what's the matter?" I said, "Father Noel just said that we had to memorize that every number multiplied or divided by zero equals zero." He said, "Well, since you know that's not true and I know that's not true, let's leave it at that." The funny thing was that Father Noel actually explained what he meant by it. He said he read somewhere that division by zero is undefined. So he defined it.

MP: *Did you have any other important influences in high school?*

Herbie Knew Everything

Banchoff: One of my big influences was Herbie Lavine. Herbie was three years older than I was. He was the son of a grocer up the street, and he was really smart. He knew everything. When I was in sixth grade he already knew how to factor polynomials, and he would teach me while working for his father. His father would say, "Herbie, you're supposed to be unloading that packing crate, not doing algebra on it." Herbie would always teach me this stuff he was learning in high school when I was in junior high. When I was in seventh grade, he told me about this neat problem involving twelve billiard balls in which one of them was either heavier or lighter than all the rest, and you were to find the "odd ball" in three weighings. I thought about it for a while. Herbie went off to college, and I went off to high school. After several years of studying the stained glass windows while in church, I worked out the solution and wrote a letter to Herbie who was at Michigan at the time. He wrote back and said it was right. He gave me another problem which was trivial, and I solved that immediately. Herbie went on to become an actuary and a professional bridge player. Herbie and his son and wife actually came and visited Brown a couple of years ago, and I took them out to lunch. I told this story to the wife and a son who couldn't believe he was the guy I idolized more than anybody else. He knew everything and taught me. He could have been a mathematician. He likes actuarial work, and he likes playing bridge. In any case, I convinced myself at that point that I could solve hard problems. Of course, that's not such a hard problem. Although I wasn't working constantly on it, it took me three years. The fact that I had been able to hold a problem in my mind for a long time and finally solve it was better in my mind than being able to solve it immediately. So by the time I graduated from high school, I had solved a tough problem and come up with a theorem and written a little paper.

I also discovered that $1^3 + 2^3 + 3^3 + 4^3 + \ldots + n^3$ equals $(1 + 2 + 3 + \ldots + n)^2$, although I didn't know how to prove it. I just noticed it. I remember at Notre Dame, when Dr. Taliaferro was lecturing about mathematical induction, he gave that identity as a

Figure 4.3 At Notre Dame in 1957, Banchoff won the Borden Prize for the highest grade point average among all freshmen.

problem. I said, "I discovered that." He said, "Now you can prove it." The thing was that I was right, and it was wonderful to discover that mathematics was something you could prove. I wasn't discouraged in high school. The teaching wasn't good, but the mathematics was good. And I was able to do anything I felt like because I was a good guy and got 100s on every test. I paid my dues in all these other clubs. I really did decide that mathematics was something that I liked and that I could do and where I could ask questions that were different from the questions that my classmates and my teachers were asking. When I got to college, I realized that was still true. That's what I liked about it. I could

have insights that were not what everybody else was having. I saw that it might get to the point that maybe you have some insight that nobody else would have. It wasn't just a question of proving things that others had already proved, but maybe you'd end up proving something that no one else would ever have proved. And that was very appealing to me. I loved the creative aspect of mathematics. I was lucky enough to realize something about the creative aspect of mathematics when I was young. These are almost trivial things, but I remember them very clearly. Each of them has later turned out to be kind of important, not that I ever did anything with the number theory, but I certainly proved theorems about cutting triangles in half, and I certainly have done things about visualizing complex functions.

Communicating—I Figured That's What Teaching Was

MP: *How did you choose mathematics and teaching as a career?*

Banchoff: My father was always a little bit uncomfortable about the fact that I seemed to be interested in teaching. But I knew I wanted to teach before I knew what I wanted to teach.

MP: *How did you know that?*

Banchoff: Teaching was big in our family. My mother was a kindergarten teacher; my aunt was a teacher; my father's mother had been a teacher. It was a respected occupation, and I liked the teachers at school. So, when I was in grammar school, I thought I'd be a high school teacher; when I was in college, I thought I'd be a college teacher. I never knew there was such a thing as a mathematician. I never knew there was such a profession. We didn't know any professionals, except for our family doctor and a couple of small town lawyers in our neighborhood. But we really didn't know anybody who had gone to college, and I didn't know about college until I went there. But I knew I wanted to be a teacher. I liked learning things, and I liked telling people. And that's what it was for me—communicating. I figured that's what teaching was. My experiences of trying

to tell other people, and having some success doing that, made me feel that I was going in the right direction. I certainly got encouragement from my mother and my aunt. My father thought I should be a banker or an engineer. He said: "You're smart enough to be a banker or an engineer; what do you want to be a teacher for?"

MP: *You could have made more money as a banker.*

Banchoff: Well, actually, he asked nicely. He asked whether I could make the kind of money as a teacher that would enable me to support a family in a comfortable way. This was a serious question.

I remember when I got my first job. My salary was $7,100 at Harvard, which was a considerable fraction of what my father was making after thirty-nine years as a payroll accountant. There I was, married with a child. At that point, he relaxed and thought it was probably all right to be a teacher.

Figure 4.4 Banchoff, the young father, with his son Tommy, wife Lynore, brother Richard, mother Ann, and father Thomas.

MP: *He must have known that Harvard was tops if you're going to be a professor.*

Banchoff: No, he wasn't impressed by the Harvard thing, but he was impressed by the $7,100. To him it meant that I could have a good life. He wanted me to be happy and to get a job that I liked. That was the important thing—a job that I enjoyed. This was from a man who didn't enjoy his job at all!

MP: *Wow!*

Banchoff: He had a terrible job. He was responsible for the payroll of about 125 people at the local gas manufacturing plant. And it was a job that nowadays could be done in about two hours a week by a semi-trained person using a program written by a high school student. But in those days, there were no computers; it had to be done by a person. Just keeping track of the week-to-week, day-to-day variations of these 125 people for a whole week entailed using big adding machines with fifteen registers. It's a familiar enough story, and unfortunately, he died before personal computers became popular. I think he would be happy if nobody else had to do what he did. He knew that it was a miserable situation. In any case, I was convinced that I wanted to be in a job that I loved. So I chose one. He was very happy about that.

MP: *Let's back up for a minute to high school, where you accumulated a rather sensational grade-point average. This all-around-guy stuff is really good in terms of gaining admission to good places. How did you decide where to go?*

Banchoff: I decided to go to a Catholic college of course. Among other things, I also was interested in theology and philosophy. I narrowed it down quickly to Georgetown and Notre Dame—Georgetown for languages and Notre Dame for mathematics. I never even considered applying to any other schools. I had visited Georgetown a few times, but I had never visited Notre Dame. I had a chance to go out to Notre Dame in my senior year because I went to the National Science Fair in Oklahoma, having won the local science fair with a math project. I had a free train trip to Oklahoma, so I went home to Trenton via Notre Dame. I walked right up the middle of the campus into the first building I saw and into the office of the only person on campus who knew who I was, the person who was in charge of scholarships. Luckily enough, I had won a National Merit Scholarship. That was the first year of Merit Scholarships, so they knew who the winners were. So I got introduced to a few people, the Dean, Fr. Charles Sheedy, and Mr. Frank O'Malley, the legendary professor of rhetoric and composition. The Dean introduced me to the mathematician Ky Fan. I said I was a high school student who had just come back from the National Science Fair with a math project. He said, "What's the project?" I said I was working on three-dimensional graphs for complex functions of a real variable. I said, "I have some models here," and he said, "You should be learning mathematics; all wrong; you shouldn't be trying to do mathematics. You should be learning mathematics." He said, "Here, read *Irrationalzahlen*. Do you read German?" I said, "Yes." He said, "Read this. Just study. Don't try to do mathematics.

You have to learn mathematics." Then I was ushered into the office of the Department Chairman, Arnold Ross. He said to me, "I hear you are interested in mathematics." I said, "I was until five minutes ago." He said, "Oh, tell me about it," in his very continental way. He listened, and then I said, "I was interested in the fourth roots of −1, and so forth, and I found a formula for it in a book. But the book was wrong. This is what it said." "So, fix it," he said, "Go to the blackboard." I said, "You mean, go to the blackboard now and do it?" He said, "Yes." So I went to the blackboard, and I figured out the correct answer. Then he smiled, and I smiled. I was a mathematician again. I didn't take calculus until I was a sophomore. I took number theory and analytic geometry as a freshman. Then modern algebra and calculus as a sophomore. This was the old days.

The Only C I Ever Got

MP: *That was very uncommon, though, wasn't it?*

Banchoff: Not at Notre Dame.

MP: *For a freshman to be taking number theory?*

Banchoff: No. That was the course. Dr. R. Catesby Taliaferro (pronounced "Tolliver") was the teacher. It wasn't a course that required algebra; it was the honors math course. If you came to Notre Dame and had the misfortune of having good SATs, you were put into the honors course whether you wanted it or not. There were four sections of

it. Donald Lewis taught one; Ky Fan taught one; Dr. Taliaferro taught one; and Richard Otter taught the other. They killed a number of the best students by having them take the course, but I thought it was absolutely wonderful. Dr. Taliaferro was a genius as a teacher—he was a mathematician, a philosopher, and a historian. He taught a marvelous course in number theory, starting with the Peano axioms and going right through Euler's inversion formulas. Then we did vector geometry and then construction of the real numbers. So it was a very fine honors course. Later I took his signature course, rational mechanics, which was given simultaneously for junior math majors and incoming graduate students.

As a sophomore I took a course in modern algebra for a year, then a graduate course in algebra as a junior, and then a third year of algebra with Hans Zassenhaus. I had a lot of algebra by the time I graduated. My analysis wasn't too strong, although I did have a full year of complex analysis with Vladimir Seidel. I also took various other courses, but the premiere course was general topology. I had heard about topology and decided I wanted to take it. My roommate, a National Merit Scholar from Dallas, Texas, Claiborne Johnson, took topology when he was a junior, and I couldn't take it when I was a junior for various reasons. He got an "A." He was the darling of Ky Fan. I took it when I was a senior. I was not only not a darling of Ky Fan, I was the butt of the class. It was made up mostly of second-year graduate students and a couple of seniors. But I didn't understand it. I didn't have the background. I didn't really know what a compact set was. I didn't do very well, and I got a C—the only C I ever got in my life. I stayed with it and got a B at the end of the second semester. But somewhere about midway in the second semester, I infuriated Ky Fan for some reason.

Every once in a while, he would turn around and ask a question. Students knew there was something that he wasn't very happy about when he did that. One day he turned around, and he said, "Banchoff, what is a set of second category?" I said, "Well, a set of the second category . . ." Well, I didn't get any further than that. "Well," he said. "Well. What do you mean 'well'? I asked you a mathematical question; 'well' is not mathematical; 'well' is English. This is not an English course, this is a mathematics course. Give a mathematical answer. What is a set of second category?" I replied, "A set of the second category is a set that cannot be expressed as a countable union of nowhere dense sets." "Right. Why didn't you say that the first time? Well!" Then he turned around and continued his lecture. So everybody in the class concluded that there was something going on between Fan and me, but it wasn't all my fault.

I took Seidel's complex analysis when I was a senior. We read Knopp's books. I didn't like the problem books because I didn't understand them, but I liked the geometry of it. At one point, I went up to Seidel and said, "These complex functions are great things—when are we going to graph them? He said, "Graph them?" I was into the fourth dimension by that time. He responded, "But you need four dimensions." I said, "Yes." He said, "Well, people tried a hundred years ago to graph these things and made models, but it didn't help." He dismissed the whole idea, but he was wrong, of course. It does help!

MP: *What were your other big academic interests at Notre Dame?*

Banchoff: I really was very much into courses in literature. I had a minor in literature and writing. I did a lot of writing in Frank O'Malley's rhetoric course. He very much wanted me to finish in three years and get into Stanford's creative writing program. I thanked him very much, but I had decided that I wasn't good enough. In my writing and literature courses, I always got the top grades, and if I didn't get top grades, I was very argumentative because I knew I was right; whereas, in mathematics I didn't always get the top grades; I got a couple of A minuses. But people explained to me exactly why I was wrong when I was wrong.

You Will Never Be a Mathematician!

MP: *Well, it's easier to see when one isn't right in mathematics.*

Banchoff: Sure. Nowadays, I look at it and realize that it's a lesser stage of psychological and moral development to want to know when you're right and when you're wrong. I didn't have the tolerance necessary to become a real writer. I knew I could be a mathematician. I did find a challenge in it, and it wasn't easy by that time. I wasn't getting top grades in mathematics, particularly in Ky Fan's class. But in my second semester

Figure 4.5 Ky Fan.

with Ky Fan, I was doing a little better, having sort of caught on to what Bourbaki is about. For some reason or other, I missed a class. I had never missed a class up to that time. I already knew the notes, but I went to him to find out something about the homework. I said, "First of all, I'd like to say that I am sorry I missed class, but I have already made up the notes." I made the mistake of telling him this in the hallway at Notre Dame. And this totally infuriated him that somehow or other I just wasn't a serious mathematics student in his mind. His summary evaluation was, spoken at the top of his lungs: "Mr. Banchoff, you will never be a mathematician, never, never!" It was loud enough that everybody was sort of leaning out their doors to see what was going on. Some of them noted that it was only Ky Fan going after Banchoff. So, that was that, and

Figure 4.6 Banchoff taught at Notre Dame in the summers of 1960 and 1961. Here Sister Cecile Rickert is asking a question about a geometry theorem.

I went off to Berkeley for graduate school. I didn't get accepted at Harvard. Lucky for me. I'm a geometer, and I never would have found that out if I had gone to Harvard.

MP: *How did you end up in differential geometry at Berkeley?*

Banchoff: The way I really got into differential geometry was by taking good notes. Since I already knew point set topology very well by my Notre Dame experience, I placed out of the point set semester and went into algebraic topology. By that time I realized that you shouldn't just study the notes of the course, you should read other books. So I started reading lots of books. But then, second semester, I sat in on Professor Ed Spanier's course, which was absolutely marvelous. I couldn't get over how wonderful it was because I already knew a lot of material. Here was somebody presenting it in such a clear way. I took very good notes, and Spanier knew that. At a certain stage, he asked me what I was going to do that next summer. I said I thought I'd probably go back and teach at Notre Dame again. (I had stayed on as a TA for Dr. Arnold Ross the first two summers after I graduated from Notre Dame.) He said, "Well, you should start doing some research. Would you like to be on my research contract?" It was a clear invitation to join him and work on some things.

Chern

That summer I attended the 1962 AMS Summer Institute on Global Geometry and Relativity at UC Santa Barbara. It was an

Figure 4.7 Banchoff came under the influence of famed geometer Shiing-shen Chern in 1962, who soon became his thesis advisor.

idea of Charles Misner, who felt that all you had to do was get a bunch of physicists and a bunch of mathematicians together, and they would realize that they were talking the same language, make the translation, and discover they were talking about the same subjects. There would be a new birth of freedom, etc. It didn't work, of course. There were marvelous lectures at the beginning of the conference by Shiing-shen Chern and Professor John Archibald Wheeler. But Chern had gone from the sum of the angles of a triangle to the Chern-Weil homomorphism in one hour. Then, the physicists had the idea that, if they could see Chern's notes for just another hour or so, they'd be able to understand it completely. He didn't have any notes. But somebody said, "I was sitting next to somebody who was taking what looked like really good notes." I took notes; I always do. Chern said, "Bring them around tomorrow morning; I'll take a look at them." At least he gave me until the morning. I worked very hard. Chern looked at them and only made a few corrections. They became the notes of his lecture. Chern's notes by T. Banchoff—wow! Chern asked if I would be taking his course in the fall and take notes. I said, "I'd be happy to." He said, "You're on Spanier's contract? You should be on my contract. I'll tell Spanier that you're my assistant now." I said, "Okay." Chern and Spanier had just come from Chicago. They were not overloaded with students at that time. A couple of years later a friend of mine went into Spanier's office to make an appointment, and Spanier said, "Before we even start talking, are you going to ask whether you can be on my contract and maybe be my student?" He said, "I was kind of hoping to lead up to that." Spanier said, "I already have had five people today who asked me. I can't take on anybody else, so save your time." When I was there as a student, he wasn't overloaded. We had marvelous opportunities. The seminars were small, and there was a lot of interchange. So I started working with Chern. He was wonderful. No matter what topic he would suggest, I sort of got into the combinatorial or polyhedral aspect of it where he wasn't at all. However, he was totally supportive and encouraging.

After a few months, Chern told me that somebody was coming to visit whom I really ought to meet. "He's like you, he thinks like you—Nicholas Kuiper." He introduced me to Kuiper and said that Kuiper had done some work in totally absolute curvature and that I should show him the model I had made in a failed thesis project. Kuiper liked it. He said, "If you like that, you might like to read some of these papers that I did and try to find some polyhedral analogs." I started looking

Figure 4.8 Nicholas Kuiper (left, with Banchoff). Chern told Banchoff to meet Kuiper because "he thinks like you."

at them. Kuiper was in town for a couple of weeks. I worked and worked and worked. I really found his work fascinating.

One day, I was folding laundry in the local laundromat. I was married by that time, and I was sketching as I was doing it. Then I said, "I'm not going to be able to prove you can't solve the problem in 5-space; you *can* do it in 5-space. Here it is!" By the time I finished folding the laundry, I had the two basic examples. I went in the next day and showed them to Kuiper who really was astounded. He looked at me and said, "You know what you have here. You have a gold mine." He said, "I'll give you six months. If you haven't written a thesis on this topic in six months, I'm going to give it to one of my students because it's too good not be done by somebody." I didn't need six months.

MP: *Did Ky Fan who was then at Santa Barbara ever come up to Berkeley?*

Banchoff: I was always terrified by the possibility that Ky Fan would end up having some conversation in Mandarin Chinese with

Shiing-shen Chern in which he mentioned that he had a student named Banchoff, and Ky Fan would yell, "Banchoff will never be a mathematician, never, never!" Chern would then call me up and say, "I'm sorry, I just heard. . . ."

As it did happen, I ran into Ky Fan at a meeting in New Orleans. I went up to him and said, "Dr. Fan, I was in one of your courses. I'm Tom Banchoff." He said, "Weren't you in my freshman course as a . . . ?" I said, "No, you're thinking of Jim Livingston." "Didn't you take the course with. . . ." I said, "No, that was Jim Wirth." He said, "Banchoff, oh yes. Oh, yes. I'm happy to see that you have developed into a mature mathematician." And I smiled.

MP: *How did you end up going from Berkeley to Harvard?*

Banchoff: I'm embarrassed to tell you how simple it was. Chern asked me, "Have you thought about what you want to do next year?" I said, "I think I'd like to go to the Institute for Advanced Study." He said, "That's a good idea. What about teaching? Would you like to teach?" I said, "Oh, yes, I really do want to teach." "Well," he said, "What about Harvard? Would you like to teach at Harvard?" I said, "Yes, I think so." He said, "I'll recommend you for a Benjamin Peirce Instructorship." A week later I got a letter offering me a Benjamin Peirce Instructorship. I went to Harvard, and, in fact, I had a wonderful teaching job the first year. I got to teach freshman calculus, which I really wanted to teach and totally enjoyed. I remember George Mackey on the phone, trying to figure out what else I would teach, and he said, "There's this one course—mathematics for nonmathematicians—would you be interested in that?" I said "What is it?" He said, "It's a general distribution course for juniors and seniors in the social sciences. You try to find topics that are interesting to these students." I said, "I'd love to teach a course like that. Is it hard to get to teach a course like that?" He said, "Would you like to teach it next year?" I said, "That would be great." There was an audible sigh on the phone, and he said to me, "The funny thing is, every year, there's actually somebody who *wants* to teach that course."

The Two-Piece Property

In part of that course, I developed for the first time a notion that turned out to be really my favorite theorem about the spherical two-piece property. An object has the spherical two-piece property if it falls into at most two pieces when you bite it. For example, a baguette does not have the two-piece property, nor does a long pretzel or anything long because you bite it so that it falls into three pieces. Biting of course, means intersecting it with a sphere. And so the question is if you intersect something with a sphere, does it fall into more than two pieces. Now a nonspherical ellipsoid can fall into three pieces, so it doesn't have the spherical two-piece property. But the sphere does, as does the hemisphere. So does a doughnut, that is, a torus of revolution. In any case, I developed this for my students in that class as I was going along trying to teach them what mathematics was as a creative process. I told them what I was working on. By the end of the second year, the second time I presented it, I really had a good idea of what the classification theorem was. I gradually understood it more and more and had to develop the methods. That investigation started the whole subject of what's called taut submanifolds. It all began with the spherical two-piece property.

MP: *How did you go from Harvard to Brown?*

Banchoff: That's also embarrassingly simple. I gave up my third year as a Benjamin Peirce Instructor to go to Amsterdam as a post-doctoral research assistant to Nico Kuiper. While I was there I got an unsolicited letter from Brown offering me an assistant professorship. The Chair, Wendell Fleming, said he had been up to Harvard, and Shlomo Sternberg had recommended me as a geometer and a good teacher. That was that.

Dean Banchoff Returns to the Classroom

MP: *You started at Brown in 1967 and became dean of students in 1971. That's quite a departure from the academic life. How did your mathematics colleagues react to your becoming dean of student affairs?*

Banchoff: I don't think anybody objected. I still taught the third-semester honors calculus course. First semester I also taught a graduate course. I also co-taught a freshman seminar modes of thought course on "growth and form in mathematics, biology, and art," which I didn't tell anyone about. When the provost asked me to be dean of

Figure 4.9 Young assistant professor Banchoff in 1970 before he was selected dean of students in 1971.

students, he told me that I wouldn't have to teach at all. I said, "I want to teach." He said, "I assumed that you would want to teach. You get paid the same whether you teach or not." I taught three courses while I was a dean. I think some of the other administrators thought it a little unusual, but nobody said anything to me about it.

MP: *But how did you actually become dean of student affairs?*

Banchoff: At the end of my first year at Brown, I was asked to be the membership chairman of the faculty club. I realized that I met the criteria. I was representing youth, and I was representing science.

MP: *Was this the "all-around guy" still at work, again?*

Banchoff: No. I think it was probably the case of somebody who was very interested in different aspects of Brown University. When I came to the faculty club for lunch every day, I'd sit at the round table where a variety of professors gathered each day telling stories. I drank in all of Brown's lore, and I'd learn about what was happening. I'd sit in on other people's courses. I sat in on a course in ethics. I was in the American Association of University Professors. At the same time, I got involved in activities throughout the university rather early. Then I was asked to be on the university housing committee. It was a heck of a lot of work about all aspects of the university, and I did it pretty well. I also played pool at the faculty club, although I wasn't very good. But I did run the table once.

MP: You were an assistant professor at this point. It's risky business to take on a nonacademic job and not pay lots of attention to research. After all, research is what leads to promotion.

Banchoff: When I was appointed to the special faculty committee charged with creating a kind of faculty senate, one of my friends took me aside and said, "This may be serious business here. We might be coming up with strong recommendations." Well, the president resigned half way through that semester, so it was serious business. I wasn't worried about getting tenure because, frankly, at that time I thought that if I didn't like it at Brown and if Brown didn't like me for some reason or other, I could go somewhere else. I was never even the slightest bit worried about political prejudice. I wasn't a rabble-rouser by any means. I've always been a very moderate, middle-of-the-road-type person. I was perceived that way, and I had no trouble talking to authority figures.

MP: How did the deanship offer actually come to you?

Banchoff: The administration wanted an inside faculty person to be the dean of student affairs, and the provost asked, "Will you do it?" It was unusual enough that I felt I should say yes. So I went home and told my wife that I had been asked to be dean. She said, "Did you say yes?" I said, "Yes. I think it's a really great opportunity. I'll see what it's like." So after a year I realized what you'd have to do in order to do it very well. I felt I'd be better as a teacher than as an administrator, and so did the students. The Brown student newspaper reported, "Dean Banchoff is returning to the classroom—a job for which he is much better suited." It was the only time that year that I totally agreed with the *Brown Daily Herald*.

MP: Your passion is in mathematics, especially the fourth dimension.

Banchoff: Correction. My passion is in teaching, and I teach mathematics. Most mathematicians do think of the fourth dimension as being rather commonplace. It is curious that it has become such a big part of my life. But it is. It's an entree for me to talk to any number of people who probably wouldn't bother to listen to me otherwise. And since I like to teach and communicate, having an entree is half the battle.

Figure 4.10 Banchoff just back from another encounter with the fourth dimension.

Banchoff Meets Captain Marvel

MP: *You said your interests in the fourth dimension started when you were very young.*

Banchoff: Way back in fifth or sixth grade. I first read about the fourth dimension in the comic book, "Captain Marvel Visits the World of Your Tomorrow." One of the panels shows a boy reporter going into a laboratory where a host guide says, "This is where our scientists are working on the seventh, eighth, and ninth dimensions." And a thought balloon goes up from the boy reporter, "I wonder what ever happened to the fourth, fifth, and sixth dimensions."

The rest of the story isn't so good. But the lead-in was really wonderful. I kept thinking about it, and I really did decide that the fourth dimension had something to do with arithmetic. I also remember a comic book

adaptation of a famous story, "The Captured Cross Section," which was written in the 1920s. It has appeared in a number of guises. In one of them, folks are terrorized by this mysterious blob that comes in and changes shape, and someone says, "It must come from the fourth dimension. See, as you put your finger through this napkin you see this?" And so they spear it, and then a scientist gets in this machine that takes him to the fourth dimension, and he comes back with his glasses all askew and his eyes wild, and he's dead because he's been in the fourth dimension. I remember that. I'd love to find that book. Once you start thinking about an idea like that, you can carry it in all sorts of different ways. So I did. I used to think about it a lot. But I really got into it when I was a freshman in high school, and by the time I was a sophomore, I really had a full-fledged theory of the Trinity.

MP: *Theory of the Trinity, as in theological terms?*

Banchoff: A sympathetic biology teacher in my school was standing outside the administration building after class, and I told him that I wanted to tell him about my theory of the Trinity. I started talking to him about this theory about how God really is a manifestation of when the fourth dimension comes into the third dimension, but all we see is a slice. So we think that it is a person like ourselves, and that's Christ. But there are two other parts we see, so that's where Trinity comes in, and so forth. I was doing this very intently, and Father Jeffrey was smiling at my seriousness about this, and said, "Why is it so important that you have your theory validated today?" I said, "Because tomorrow I'm going to be sixteen years old." I was interested in it in high school, and I still remember boring my friends crazy in pizza parlors with my napkin, doodling. I talked to my pastor about my ideas. He's retired now. He's seventy-five or so. He said, "Well, Tom, I'd have to say I thought you were probably going to end up in a heresy, but you seem to have done all right." As a sophomore at Notre Dame, I did my theology term paper on the fourth dimension and the Trinity.

MP: *Let's talk about the reactions that you get from your listeners and your readers when you discuss the fourth dimension. It's clearly an exotic topic.*

Banchoff: It's kind of mysterious and jazzy, and people think of it as kind of offbeat. They come to find out about it out of curiosity. Most of the people I talk to are kind of bemused by the experience. They like the visuals a lot, especially the computer graphics films we started making in 1968. It's a topic that makes people feel challenged. It's something they hadn't thought about before, and it stretches their minds in some way. Many people say, "I didn't quite understand it all, but if I think about it a little bit more, I think I'll get it." I say, "That's okay. I don't understand it all myself." It would be terrible if they understood it. Well, it's funny, because in a sense you do want to leave people with the idea that there are many different kinds of fourth dimension. That's the biggest message. As I say in my talk, "I started out this talk by saying what's a one-word definition

Figure 4.11 Banchoff studying a hypercube. He first became interested in the fourth dimension as a fifth grade student.

of the fourth dimension? What would it be?" Many people say "time." Why would that be the wrong answer? Because it's the wrong question or a wrong question. Time isn't *the* fourth dimension; it's *a* fourth dimension. And it's something as prosaic as making an appointment with somebody on Third Avenue, Fourth Street, the fifth floor at 6 p.m. You could write 3–4–5–6 in your appointment book. But if you talk about time as *a* fourth dimension, not *the* fourth dimension,

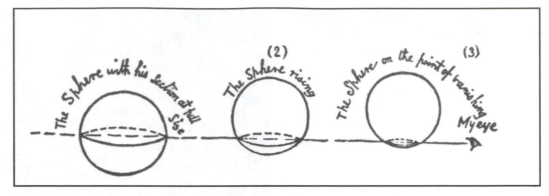

The Sphere with his section at full size (2) The Sphere rising (3) The Sphere on the point of transiting My eye

Figure 4.12 As A Sphere passes through Flatland, the two-dimensional section changes, starting as a point, reaching maximal extent as a circular section, and then reducing to a point as it leaves Flatland.

Flatland

In 1884, a British schoolmaster, Edwin Abbott Abbott, wrote the classic introduction to the dimensional analogy. His small book *Flatland* is narrated by A Square, living on a two-dimensional flat universe and as incapable of comprehending our geometry as we are when we try to conceptualize a fourth spatial dimension. We are invited to empathize with the experiences of A Square, first of all in his two-dimensional world, a social satire on Victorian England, and then as he is confronted with a visitation by a being from a higher dimension, A Sphere, who passes through his universe, giving to the two-dimensional onlooker the impression of a growing and changing circular figure. This experience challenges A Square to rethink all that he had taken for granted about the nature of reality. Analogously, we are challenged to imagine the experience of being visited by beings from a fourth spatial dimension.

then that puts people off the defensive a little bit. Then they may be willing to think about something else as being many-dimensional. And then you go from there. The idea of slicing a hypercube seems like a reasonable thing to do even though nobody has ever really had to do it up to that time. In our hypercube movie, we stayed with the really basic, fundamental, and elementary stuff. It's hard enough. The things that are raised there are central—what we're actually experiencing when we see the rotations of the four-dimensional cube, what it means to see something rotated, and to understand what rotate means, and what perspective means. These are questions that we should be asking about our ordinary experiences. These things make people look at their own experience in a different way. We don't have a lot of converts running out to do four-dimensional stuff, but at the end of my courses and even at the end of my lectures, I have people doing three-dimensional things differently than they did before. That's sort of being conscious of the dimensionality of experience. It's a kind of a paradigm of experience as a way of organizing what we do and feel and having to come to terms with it. Visualization is very useful if you can separate things into different parameters and translate them into dimensions and translate that into some geometric form that you can look at, think about, and manipulate. Then you have a better chance of making sense out of whatever complicated thing you're trying to deal with. That's, I guess, the basic message.

The Only Thing I'm Good At

MP: *Maybe you're luckier than a lot of mathematicians. You've had the good fortune of choosing an area that is imaginable for a wide group of people.*

Banchoff: Well, I didn't have much choice. It's the only thing I'm good at. I simply followed my own lights.

I feel lucky because I do have some phenomena to work with. The phenomena I study are things that I can present to people, and a number of people can appreciate them quite independently of understanding why I care about them, how I created them, or what they're for. They are beautiful, after all. They are intriguing. All my life I've worked with polyhedra, I've worked with drawings, I've worked with models, and now, in the last twenty years, with computer graphics. There's no question about the fact that the images that come out are fascinating. It's sort of a different kind of fascination than fractals. Fractals are fascinating because you keep going and going and going—you never run out. But I work with things like hypercubes, which are so simple that you can describe them in a line; and yet you can watch the film "The Hypercube" three hundred or four hundred times and always see something new.

The inexhaustibility is not because the object itself is infinitely complicated; it's just that the set of relationships that are available there is infinite. There are infinitely many views that you can take of an object, and there just seems to be no limit to the ways of combining these insights. You show it to people, and people react to these things at many different levels. You're sharing, and you're telling them some of the things that it means to you. You're not telling them, "This is insight number 397 or number 398." You're just saying some words and letting them organize their own insights. And that's rather nice. It's a very pleasant time to go up there and parade some of your mathematical children in front of people, and they applaud, or at least they smile.

Most mathematicians probably aren't so lucky. I think I mentioned this before: I'm one of the few people who can actually tell people what his doctoral thesis was about. Most mathematicians have no chance to tell. Of course, after I tell what my thesis is about, I have to explain to them why it's not trivial.

MP: *Your work certainly gets a lot of coverage in the popular press.*

Banchoff: I remember the very first time there was an article about our work, a *Washington Post* article on the fourth dimension. One of my mathematician friends, who was in Washington for the mathematics meetings, came up to me and said, "It was a nice article, but I don't see why people are interested in that. Our work on norms on Banach spaces is much more interesting." I didn't say anything, except that there's no accounting

Figure 4.13 Banchoff signing copies of his book *Beyond the Third Dimension*.

for taste. What can I say? Some people are fascinated by this business, while others ask "What does this have to do with the price of tomatoes?"

MP: *Who asked?*

Banchoff: Tom Zito was the *Washington Post* reporter who interviewed me. And so I started talking about using dimensional representation for brain research and for various economic models. Real mathematicians know you don't have to answer that question about what math research is "good for." You can say, "I don't have to answer that question." It's a right answer because, as a matter of fact, I am excited. I got a call just a month ago from Tom Webb who's a geologist-paleoclimatologist in a building next to mine at Brown. We'd been looking for years for a chance to collaborate on something. His data were four and more dimensional. His data have to be understood visually, and I think they can be understood from a four-dimensional point of view much better than any other point of view, and he thinks so too. He borrowed the Hypercube film to show to a bunch of geophysicists out in Colorado to try to convince them that they have to forget about the two-dimensional representations and go to families of three-dimensional representations sliced various ways—honest to God, four-dimensional stuff. His data sets are all four-dimensional in the sense that you have latitude and longitude to locate a position of a core sample from a dried-up lake bed. A core sample is a very nice way of translating time into space because the further down you go, the further back you go in time. And then, when you're down there, you can look at something like the density of spruce pollen and use that information to tell you what was happening in the neighborhood of the core at that time. There's enough experience spread over a two-dimensional plane that you can make a whole family of time series. And you really develop a four-dimensional continuum of two-space, one time and one density. Seeing how these things interact, sliced in long transects, river beds and so forth, you get three-dimensional slices in different ways, not just trivial ways but ways that respect the geometry and the geology of the situation.

The lucky thing about collaborations like this is that I get to meet all these great people outside mathematics. People come to me with problems. In one way or another, I interact with lots of people. New problems get suggested. And I follow them up. Not all of them. You get a dozen nibbles a year and maybe one or two get beyond the second phone call.

MP: *Give me a little sampling of these people who are calling you up.*

Banchoff: I worked with an anthropologist who has two projects that involve mathematics. One has to do with the ranging behavior of aboriginal tribes, hunter-gatherers, who operate in places where there are no large predators. So wide circles represent their range from the camp, and then after a while they move to another place—not too far away, but not with too much overlap from the original camp. So the overlapping behavior of circle patterns over time is something that we happen to be discussing when we are also talking about the geometry of circles in a plane.

The same fellow, Richard Gould, was doing underwater archaeology. They were studying a site in the Caribbean where there had been a number of ships that had sunk and then fill was brought in on top of them. They're trying to excavate them, but it's very difficult to excavate things under water because, if you take them out of water, they rot. So you really have to be able to work at the underwater level. So what you have is a slice of a large hull. And by looking at that slice, they'd like to predict exactly what the aspect of the rest of the hull is. So it's very geometric in the sense that you're trying to do a reconstruction from a small amount of data. That's very exciting.

MP: *How about artists?*

Banchoff: Artists, writers, dancers.

MP: *Dancers?*

Banchoff: I cooperated with Julie Strandberg, a choreographer who did a wonderful piece a couple of years ago: DIMENSIONS. It was performed in New York and twice at Brown. It was absolutely great. It was so thrilling. I cry every time I see it. It's a sad story (chuckling). There's this square, you see. It starts out with all these poor creatures who are limited to a two-dimensional space. They dance with their backs up against the wall. They can't even pass each other unless they go over the top. They can do cartwheels, but they can't do somersaults. Then if they go into Egyptian mode, they can sidle back and fourth, but they can't really pass one another. But after we see a lot of these contortions and so forth, there's a visitation from a being from a higher dimension on the screen. And everybody runs off except one square. She stays around to find out what's going on and to have a chance to see people operating in the third dimension. There is a *pas de deux* that gives me the chills just to see it. It's a marvelous thing watching professional dancers do this with the square watching; the square is wearing gray leotards and these people are the rainbow. And then finally she very timidly comes up and is welcomed and is pulled in between the two of them. Finally they take her out and liberate her into the third dimension. But in the end she has to return to the plane. It's very sad.

The Joy of Teaching . . . Undergraduates

MP: *Brown focuses on undergraduates. Was that part of the original attraction of Brown for you?*

Banchoff: I like teaching the undergraduate courses more than I like teaching the graduate courses.

MP: *Why?*

Banchoff: I always feel a great sense of duty when doing the graduate courses that I have to cover lots of material whether I particularly find it interesting or not. I had about fifteen students this year, and not all of them were very geometric. I thought I had to make it possible for them to get something out of it even if they weren't geometric. The geometric ones I had no problems with, and they had no problems. They got involved in it, and I'll have one or two thesis students out of it, I'm sure. I felt I just couldn't produce stuff that's formal enough even though I'm fairly formal myself. Differential geometry is even harder because there's a lot of technical stuff you have to do in the first-year course.

MP: *Do you really think that's true that you must get through all that technical stuff? Are you worried that in the process you may be*

damaging attitudes? *Can the joy of doing the subject be lost by pushing people through all the techniques?*

Banchoff: I think it is important to develop a common joy of the subject. I think it's important for each mathematician to let her or his own joy show through. My strength is not in teaching a basic, technical first-year graduate course. I can do it. But as it happens, I very much like doing it the way we did it in algebraic topology. I taught the first semester in which I presented quite a bit of the motivational stuff, geometric stuff, if you will, with a lot of good problems. In the second semester a very talented assistant professor who likes the formal stuff very much continued with the same class. I think it's very important for the students to see both. I think it's terrible if they see only the formal thing, but I think it would be equally terrible or even more terrible if they saw only the intuitive stuff, because in this day and age you have to be able to do the formal stuff.

MP: *But the formal technique generally will come more easily if the intuition is well developed.*

Banchoff: I think you're right. I don't know of any first-rate researchers who don't have a very well-developed intuition. And I know that it's easy to be judgmental, but I've watched a number of situations in which people sort of tried to follow the coattails of famous mathematicians and never made it anywhere near the stature of the person they were trying to emulate. They were trying to talk the way the person talked, but what they were missing was the fact that that person

Figure 4.14 Student Deborah Feinberg learning Stokes's theorem from Banchoff

had a couple dozen years of experience that they never articulated, but it was behind all the right insights that they had. Just talking like somebody doesn't mean you share [that person's] insights.

One of the academic sins is coming up with good results and not doing that last bit of really hard work. It's too easy to get up and tell your contemporaries about them. Most of the people who are working in a field know my definition of the normal Euler class and exactly what the properties are. But the only people who know that are the people who knew me. The people who are interested in this particular topic generally do know me, so that's not the problem. The problem really is that I'm not even writing for a larger

community outside those individual people whom I run into, and I'm certainly not writing outside of time and space. Writing articles is not just for people of this generation. There are people who will possibly be interested in what I write a hundred years from now. Unless I'm scrupulous enough to write it down seriously, nobody a hundred years from now is going to be able to read it whether they want to or not. So, part of the responsibility is there.

MP: *How important is the geometric approach to you?*

Banchoff: For me, a lot of times, when I hear a result that sounds very geometric, frequently I can't understand the proof at all.

If it's really geometric, I want to try to find my own proof. I want to see what it really is. That's happened to me a lot.

MP: *Wouldn't it be easier, to maybe learn some of this other stuff?*

Banchoff: No, no.

MP: *Why not? Wouldn't that enrich you? Wouldn't that give you even more power?*

Banchoff: The fact that I've always tried to find the geometric content of theorems, say, about characteristic classes has enabled me to discover some things that other people passed by because they knew too much. They didn't care about these problems. I wanted to find out how it worked, for example; how it worked for something else. And I would stay with examples until I understood something, and then I would prove something, and then occasionally something new would come out of it. It's almost a quirk to try to find what I call the geometric content of theorems. Occasionally, it leads to a new insight as it did several times. Again, the sense of creativity comes with a feeling of uniqueness, not that you were the first person to reach the finish line in a race where there were a lot of other people running together and somebody would have gotten there ultimately, but maybe you find another path through the woods that nobody else would have taken.

MP: *You've nailed some nice problems. What can you compare it to in the experience of others that they might relate to? Is it like winning a race? Is it like pumping a lot of iron—more iron than you ever imagined you could pump?*

Banchoff: I think it's more like taking a walk with a bunch of friends, up on a mountain trail. Maybe other people have gone off in another direction, either ahead of you or coming up behind you. You see something that looks kind of interesting and for some reason or other you come upon a particular rock and look out and see an extraordinarily beautiful landscape, a beautiful expanse, a beautiful sunset, and you call your friends, and they come up and they agree. That's really beautiful. It might mean that everybody will decide to go off in that direction.

MP: *But how long does that last? Do these experiences, these special experiences, come back from time to time? Do you think about them? Can you remember the feelings?*

Banchoff: Yes. I can remember some of the feelings from the first time I discovered a polyhedral surface in five-dimensional space that had the two-piece property.

MP: *You still can remember that?*

Banchoff: Oh yes, that was back in that laundromat on the east side of Berkeley. I was folding laundry, and there it was.

The essence of the creative act is putting ideas together in a way that nobody ever has before and maybe nobody is likely ever to do again. You come up with something truly original, not just before other people do, but quite independently of what anybody else ever might do. Maybe I'm wrong. You can't prove that somebody else wouldn't have come along and done it exactly that way sometime later on. But it doesn't have very much meaning to say that. I have a few things I'm proud of that I can show people, much in the same way that my friends who are artists would come in and show their favorite sketch or their favorite oil painting or their favorite ceramic sculpture. I really envy people who can actually hang something up on the wall. It's kind of hard to hang up a theorem.

Dalí Meets "The Mathematician"

MP: *Who is the most unusual person you've ever met?*

Banchoff: Salvador Dalí. He's certainly the most outrageous person I've ever met.

MP: *What was the occasion?*

Banchoff: He called us. The *Washington Post* had written this article about our work back in 1975. They had included a picture of me holding my own folded hypercube, hair sticking out, Daliesque. And as an insert in the back of the picture, they had a picture of Dalí's Corpus Hypercubicus, body on the hypercube, the famous surrealist painting. When I came back to Rhode Island, my computer science collaborator, Charles Strauss, said to me that he was very surprised to see that they had included that picture in the paper because, although Dalí is well known for his love of publicity, he loves it on his own terms, and he was surprised that the *Washington Post* hadn't gotten Dalí's permission to use this. A couple of weeks later, when I came back from class, there was a note saying: "Call so-and-so in New York representing Salvador Dalí." I said to Charlie, "What do you think?" And he said, "Well, it's

Figure 4.15 The mathematician Banchoff with the artist Salvador Dalí.

either a hoax or a lawsuit." I called up, and a woman answered the phone and said, "Oh, yes, Señor Dalí would like you to come to New York and meet him." So, I said to Charlie, "What do you think?" And he said, "The worst we can get out of it is a good story." So we went down to New York City, and we got not just one but a whole set of good stories. We met Dalí at the St. Regis Hotel in the cocktail lounge. He was holding court. All sorts of strange people would come up and bow to him in a sense and be asked to sit in some large circle around him.

MP: *How were you introduced?*

Banchoff: As the "Mathematician." Actually, not then. But he had called us, and he wanted to know about what we were doing. He was very curious about stereoscopic oil painting and Fresnel lenses and various other ways of displaying it. We hit it off very well. Since he was picking my brain, I thought I would pick his too, so I asked him about his painting. I showed him my model of the hypercube. He looked at it and said, "I may have this." It wasn't a question; it was sort of a regal statement. I said, "Well, yes, you may have it."

MP: *What did he mean by that?*

Banchoff: He wanted it for his new museum in Catalonia. Later on, I saw the model now exists; there's a model of it in his museum in Spain. No attribution, mind you. I asked him where he got the idea. He said, "Ramon Llull," or something like that. I don't want to try to repeat his accent because in public he would speak with kind of a mixture of French and Spanish and English and Catalan. Anyway, I finally realized who he meant. I said, "Ramon Llull, do you mean the famous twelfth/thirteenth century polymath?" He said, "Yes, yes. You know Ramon Llull?" I said, "Well, yes, strangely enough I do." I sat in on a course taught by a friend of a friend on the history of the Church in Spain and Italy in the twelfth and thirteenth centuries. I took it because it was something different from math and something I could do instead of math. It was a break, like reading detective novels. Well, the teacher mentioned Llull, so I read some of his work, kind of a fatuous work in which he proved all the major theorems of antiquity, and various other things. Dalí was very impressed that I knew who Llull was. I was very impressed that Dalí knew who Llull was. So, he invited us down. My wife and I came down the following week because he wanted to see our films, and we showed our films to him and his wife and his managers. Every spring he would invite us to come back down and bring our videotapes and films and slides, and he'd show us what he was doing. That went on for about ten years.

Five

Leon Bankoff

Gerald L. Alexanderson

Leon Bankoff was born in New York City on December 13, 1908. Thus, in the spring of 1986, he became one of the few who were able to witness for the second time the appearance of Halley's Comet. Although he had an early interest in mathematics, he did not choose the subject for a career, so after college at CCNY and dental school at NYU and USC, he became a prominent Los Angeles dentist whose list of patients includes rock stars and other luminaries of the Beverly Hills-Hollywood scene. With a resurgence of his interest in mathematics in the late 1940s, he went on to publish regularly in mathematics and to edit the Problems Section of the *Pi Mu Epsilon Journal*. He numbered among his friends illustrious mathematicians from around the world.

MP: *Let's start at the beginning. Were you born interested in mathematics? Or did that interest develop later?*

Bankoff: I must have been about five when I became interested in mathematics.

MP: *Numbers? Or figures?*

Bankoff: Numbers. My mother gave me a problem, the old two-jug problem about measuring out four gallons of a liquid using containers that could hold only five and three. I recall that I solved the problem, and I got a pat on the back for it.

MP: *Did your parents have a background in mathematics that would have caused them to encourage you in the field?*

Bankoff: My parents were not professional people and were not scholars in the sense that they possessed great erudition. Yet, I could say that they were scholarly, always eager to learn and to learn quickly. My mother, an immigrant from Riga, and my father, who came from Pinsk, came to Manhattan at the turn of the century, where they met and married in January 1908. I did not know my grandparents on either side since they all remained in Europe. By the time I arrived in December of 1908 my parents had already learned to speak, read,

Figure 5.1 Bankoff, dressed as a Russian diplomat, Le Mans, 1970.

and write English. My father plied his trade as a manufacturer of custom hand-made cigars and continued his craft throughout his life, extending his activities as a merchant of related tobacco products and accessories. My mother, a talented seamstress, quickly found herself swamped with more work than she could handle, thus forcing herself to expand by hiring a dozen seamstresses. She became a well-known Fifth Avenue couturière.

Music at an Early Age

Both of my parents were music buffs, mostly in opera, and my earliest recollections involve open-air concerts in Central Park and the Lewisohn Stadium, plus almost incessant uproars emanating from our wind-up Victrola. My mother fostered my early interest in music and kept me busy with all sorts of math problems and puzzles. My father escaped early evening boredom by teaching me chess and checkers, pushing me way ahead in arithmetic, entertaining me with magic tricks and calligraphic flourishes. They provided me with all the books I wanted, and I didn't have to be coaxed to read them.

MP: *Later on was there anyone in particular who was influential in developing your interest in mathematics, any high school teacher, any friend or relative?*

Bankoff: My father taught me subtraction, multiplication, and division before I entered school. By the time I entered the first grade, I was a little overqualified for it, so I skipped two grades. As a result I graduated from elementary school—which ordinarily would have taken eight years—after six years. I had some teachers I remember. There was a man named Mr. Lawyer, who presented the Pythagorean theorem to these young kids who were not really ripe for it. I was terribly impressed with the fact that three squared plus four squared equals five squared.

MP: *It's a beautiful theorem!*

Bankoff: It really is. That theorem brightened my young brain; the appeal was geometric and number theoretic. I used to amaze my young friends by being able to generate Pythagorean triples by a simple, almost naive method. I would start with any odd number, multiply it by itself and divide the result by

2. The result, of course, was not an integer, but by adding ½ and subtracting ½ I could obtain two integers, which, together with the original odd number, produced the desired triple. I must confess that I didn't know how to generate triangles such as the 8–15–17, 28–45–53, etc. That came later.

I must add that another one of my great talents was the construction of magic squares. I learned the trick at a rather early age and used it to mystify anyone who needed to be mystified, mainly my preteenage friends.

Before entering high school I used to browse around in bookshops, and I picked up Wells's *Algebra* where I found you could find out things about apples and oranges by using *x*'s and *y*'s. I went through the entire book before I had any formal classes in algebra. I spent a summer vacation solving all the problems in the book, so by the time I got to school I didn't have to do any homework. All I had to do was copy my notes.

MP: *So with this head start in mathematics, did you major in mathematics?*

Bankoff: No, I did not. I majored in general science. I took whatever mathematics I was required to take: algebra, Euclidean geometry, trigonometry, and some solid geometry, believe or it not. And in college, it was analytic geometry and calculus, and that's where my formal mathematical education terminated.

Dentists—"I Liked Their Lifestyle"

MP: *Did you go directly from college to dental school?*

Bankoff: Yes.

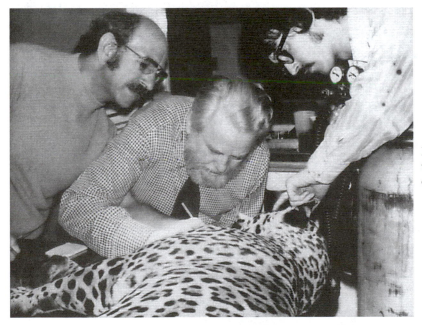

Figure 5.2 Bankoff, dentist to Hollywood stars and the occasional leopard.

MP: *Why dentistry?*

Bankoff: That's a question that is frequently asked. It's not hard to answer really, because I entered college at the age of fifteen and graduated at eighteen. I didn't know what career I wanted, really, but I very cleverly managed my courses so that I would qualify for anything I chose at the end of my college education. I could go into medicine, dentistry, engineering, pharmacy, or teaching. The reason I chose dentistry was that I knew quite a few dentists. I liked what they did. I liked their lifestyle. At that time physicians used to make house calls at two o'clock in the morning. I didn't care for that sort of life, and for that reason medicine never attracted me. As far as all the other professions that were open, accounting, and so forth, they never appealed to me. But dentistry did and for another very good reason. I was always

Figure 5.3 Leon Bankoff, the student, drawn by Frank Netter, the famed medical illustrator, 1927.

interested in sculpture. I noticed that dentistry was an outlet for my aptitudes for painting, sculpture, and so forth.

MP: *Where did you go to college?*

Bankoff: CCNY, followed by dental school at NYU for two years, then I transferred and took my last two years at USC.

MP: *So that was the move to Los Angeles.*

Bankoff: My family moved to California, and I moved along with them.

MP: *Between the time that you finished calculus and a certain number of years later, you were not doing any mathematics.*

Bankoff: I used to dabble with various puzzles, the Sam Loyd and Dudeney stuff.

MP: *What got you started on real mathematics? Was there a single event or a single person?*

A College of Magic

Bankoff: Yes. Among my patients, way back in the forties, was a magician, and he told me that he was studying in a college of magic run by a magician named Ben Chavez. Chavez had a college of magic that was the only one authorized by the United States government to give instruction in magic to GIs coming out of the service, people interested in that as a profession. I wasn't a GI, but I was interested in magic, and I enrolled in this school, took the course, and got my diploma in prestidigitation. Naturally, as I do in anything I'm interested in, I accumulated a library. In that library of magic books were a number of periodicals. One of them was called *The Bat*. It had among other things a column called the Puzzle Corner to which various people contributed problems. Among them was Martin Gardner. This was in the middle forties, before he became famous for his column in *Scientific American*. In one issue there was a problem, and they offered for the best solution a prize of a bound volume of the previous issues of *The Bat*. I won the contest. I got my prize, and in reading subsequent issues, I kept seeing references to a journal called *Mathematics Magazine*. I went to the newsstand but I couldn't find anything called *Mathematics Magazine*. But then I discovered that there was such a magazine, published locally in Los Angeles. I subscribed, tackled some of the problems, and found that practically all of my contributions were accepted and published by the problems editor—at that time, Charles W. Trigg.

MP: *Ah, Trigg enters the scene.*

Bankoff: We became acquainted. I would solve a problem and tear down to Los Angeles City College, put it on his desk and say, "Look at this!" He put me onto means of acquiring more knowledge of mathematics. He told me to buy Nathan Altshiller-Court's book on college geometry, various books on number theory—Oystein Øre, Uspensky and Heaslet—and books on trigonometry. I gradually started accumulating a library by searching bookstores for second-hand books. I wandered into a shop on Main

Figure 5.4 Bankoff (right) with Charles W. Trigg, San Diego, 1971 (reading Pedoe's book on geometry).

Street in Los Angeles one day and asked whether they had any mathematics books. They said they had some down in the basement, and I was surprised to find there about a dozen books called the *Ladies' and Gentlemen's Diary*. I bought them, of course. Later I had occasion to consult an issue of the *Diary* that I didn't have, and I discovered that the William Andrews Clark Library (part of UCLA) had it. So I went down there, and to my surprise I found that issues from my run were not in the Clark Library and the other way around. I told the librarian I would either have to buy his issues or will him my copies. But this served as a great impetus to my inquisitiveness in mathematics. I raided the bookshop of Zeitlin & Ver Brugge (the red barn on La Cienega), now gone, and I was able to get some rare books like Pappus in Latin and some Archimedes. I gradually accumulated an immense library. I bought everything that came along that was of interest, not only geometry but practically every area of mathematics.

MP: *I remember Jake Zeitlin's bookshop very well. I recall that in London I was once looking for some obscure work in the history of mathematics, and I was told there that my best bet for something like that would be to try Zeitlin in Los Angeles. It went against the conventional wisdom about locating scholarly books.*

Bankoff: I knew Jake Zeitlin very well. I had lunch with him just two weeks before he died. We tried out a new restaurant and that night both of us took ill.

MP: *That sounds ominous—more than a coincidence.*

Bankoff: Well, not really. He went to the hospital for a heart condition, and I had a gallbladder attack.

Soon after I met Jake he showed me a rare copy of Lewis Carroll's *Pillow Problems,* which he wouldn't sell. It was so precious he wanted to keep it himself. But he lent it to me, and I copied it by hand, word for word, even copying the style of the printing. Later, when Dover came out with an edition of it, Jake called me up and asked me whether I wanted to buy it!

Anyway, I might mention that when I took my mathematics in school, I had no trouble with it. For instance, before calculus I was on summer vacation. Since I was old enough then to start going out with girls, I didn't want to waste my evenings doing homework. So that summer I solved practically every problem in the book. During the year, when I got assignments all I had to do was copy out the solutions. Then I had my time free to pursue my social life.

MP: *You learned calculus in college in 1926. And you encountered a magician, who indirectly got you interested in mathematics, in roughly the mid-1940s. So there was a twenty-year period when you were not doing mathematics. You were, of course, doing dentistry. What else were you doing to occupy your time? Music?*

Bankoff: I had been playing the piano since the age of twelve. I used to play the violin, but I gave that up when I went to college because I had started smoking, and I couldn't tolerate the idea of ashes dropping into the f-holes. I became a chain smoker, and I always played the piano smoking a cigarette.

MP: *It provided a spot for the ashtray.*

Bankoff: No, at one time I had an ashtray constructed that I wore around my neck. A friend of mine made it so I could play the piano without any regard for the ashes. But I am glad to say that in 1969 I gave up smoking.

MP: *So you played the piano, violin, and, I understand, somewhat later the classical guitar. And we've heard about sculpture.*

Bankoff: I used to spend my Sundays taking the ferry across to New Jersey and painting

Figure 5.5 Bankoff, the classical guitarist, 1978.

Salesmen would come in to sell me dental equipment, and we sat and played chess.

MP: *You sound a little bit wistful about the ease of life then with a lot of time to spend on your interests outside dentistry. I assume that life has gotten busier.*

Bankoff: It has. I'm very intense.

MP: *So all of these things were going on and then, mathematics. Since you got interested in mathematics, you have published quite a bit. Do you have any idea how many papers?*

Bankoff: Not terribly many papers. Maybe about a dozen. My chief activity has been in posing and solving problems.

with a cousin of mine who was quite a good painter and later owned a gallery in New York. Sculpture always interested me. But most of my time was spent as an equestrian. I belonged to a riding troop, and we used to do all kinds of military maneuvers on horseback. I enjoyed that very much. And I spent a lot of time ice-dancing. There used to be numerous ice rinks in Los Angeles. Sonja Henie owned one in Westwood before Westwood was built up. There was a Pan-Pacific rink and one in Pasadena. I used to go Sundays with my daughters after they grew up. They all received an ice-skating education, figure skating. I also have been very interested in calligraphy. And I played chess a great deal. In the early days of my practice, I cannot say I was terribly busy. I started practice in 1932, in the depths of the Depression—there was plenty of time to play chess.

An Erdős Number of 1

MP: *You seem to know a good many of the important mathematicians of the mid-twentieth century. You have traveled in good mathematical circles. Not all that many mathematicians one meets, for example, have an Erdős number of 1. But you have co-authored a paper with Erdős. Have you had other co-authors?*

Bankoff: Well, there's Charles Trigg, of course, Jack Garfunkel, and Murray Klamkin. I can trace my acquaintanceship with all those mathematicians to Charles Trigg. He encouraged me to write to Victor Thébault of France. There was a problem published in the early fifties in *Mathematics Magazine* about determining the radius of a circle inscribed in the figure known as the shoemaker's knife of Archimedes. I became fascinated with this problem, and I sent in three separate solutions, which were published. Among the other solutions published were one by Howard Eves, one by Klamkin, as well as others. I asked Charles Trigg where I could learn more about this particular figure, and he told me that there was a French mathematician who was at that time about the same age I am now, who had written extensively about the shoemaker's knife of Archimedes. Thébault was very well known in those days, and, in fact, I contributed an article to the *American Mathematical Monthly* about him, along with Starke, Eves, and others. Anyway, when I found out about his work I wrote to Thébault and discussed this matter, and he sent me numerous reprints of his. I made some discoveries of my own, which I communicated to him. And after a short time it was decided that we would compile a ten-chapter book on this one subject, where he would contribute five chapters and I would contribute five.

Figure 5.6 Victor Thébault (left) and Bankoff, Tennie, France, 1958.

MP: *Had you met him at this point?*

Bankoff: No. The correspondence took place very rapidly because in those days when you sent an airmail letter to France it would arrive in three days. If I wrote on Sunday, I would have a response the following Sunday.

MP: *That's because the planes were a lot slower in those days!*

Bankoff: And the mails were a lot faster!

A Visit to Tennie

Every summer he invited me out to visit him in his little chateau which he called "Le Paradis," which was located in Tennie, a small village—really not a village at all. It's not even large enough to be called a hamlet. It had a population of about two hundred possibly, if that many. Thébault's telephone number was Tennie 2; the mayor had Tennie 1. At any rate, after a few years I decided something had to be done. I had developed a lot of material on the golden arbelos (shoemaker's knife) and various ramifications involving Pythagorean triangles, rational triangles connected with the arbelos, and in 1958 I wrote that I would be there. I went to Paris, then took a train to Le Mans, where Thébault was waiting for me with a young lady named Francine Laloue, who served as an interpreter. He hired a taxi to take us twenty kilometers away to his little hamlet, Tennie. We became acquainted. I showed him what I had done. He thought it was great stuff. After a visit of about three or four days, I left. I was supposed to continue work on this manuscript, but I got bogged down with so many other things, mainly dentistry, that the thing was delayed and delayed. Even to this day, twenty-eight years later, the manuscript is still residing in my closet, unfinished.

MP: *But one of these days. . . .*

Bankoff: One of these days I expect to retire and get this manuscript to the printers.

About five years ago Martin Gardner wrote about this unpublished manuscript in the *Scientific American*. I have had numerous inquiries about when it will be published. Someday I hope to finish the job. If I don't live long enough to do it, I have asked Clayton Dodge to get the material I have and to get it into print.

I admired Thébault's work greatly. His work on the shoemaker's knife was really incidental. He did a great deal of work on arithmetical recreations, intense work on number theory. In fact one of the first problems I solved was a problem that Thébault had proposed, to find a cube plus a square that is equal to 2,000,000. He was also very deeply into the geometry of the tetrahedron.

I wouldn't say Thébault was penurious but he used to write on the back of calendars. He was retired when I met him. He did mathematics. He showed me around his estate and also suggested that I buy his property. I had no use for it, but I am sorry now I didn't. Ten years after he died his widow died. I used to make annual visits to her. Shortly after he died I was ushered into his library, which she kept as a shrine. His desk was left exactly as he had left it, pen here, eyeglasses there, letter scale over there. And it was rather eerie

to see there a letter of mine on the desk that he hadn't replied to. After his wife died I was invited by one of his sons to help myself to some of the items in Thébault's library. So I have a complete run of the *Educational Times* that belonged to him. And I have his moderately large run of the *Mathematical Gazette*. He was fond of making jokes, and he smoked a great deal, as I did. I remember he once went to get champagne which he kept in a cupboard that looked like a grandfather's clock. But it wasn't really a clock. Before he retired he was a director of insurance, sort of an actuary. And for a while he taught school. But when I knew him he did mathematics.

MP: *Your interpreter on that visit has a special role now for you. How long after meeting her were you married?*

Bankoff: Eight years. I'll tell you how that happened. In 1963, five years after my visit to

Figure 5.7 Leon and Francine Bankoff in a restaurant in Paris, 1958.

Thébault, this young lady was invited by the University of California, Berkeley, to come and set up audio-visual language instruction in which she was expert. And when she arrived here she communicated with me, and we reestablished our acquaintance. In 1965, because of visa regulations, she had to go back to France for a year. She came back again in 1966, and we were married that year.

MP: You are largely associated with geometry in people's minds, although you mentioned to me that you are really interested in inequalities. Have you published as much in inequalities? Why do we associate your name more with geometry?

"I Disagreed with the Published Solution— I Found a Counterexample"

Bankoff: Well, geometry is more visible, for one thing. Inequalities probably do not attract as much attention. But my interest in inequalities started a long time ago. My first really good achievement was with a problem that was proposed by Thébault, by coincidence. He proposed a problem concerning some trigonometric inequality, and I sent in a solution. My solution was not used, but I had to disagree with the published solution for the simple reason that I found a counterexample. I communicated with Howard Eves, who was at the time the Problems Editor of the *Monthly,* and, sure enough, he said I was right. Would I offer a correction? So that too was published in the *Monthly.*

Figure 5.8 Bankoff with Sol Golomb at Cambridge University, 1987.

MP: You mentioned that this work appears in a standard work on inequalities.

Bankoff: Yes, Bottema's book on geometric inequalities. When I say geometric inequalities, I always associate geometry with trigonometry. Practically all of my geometrical inequalities have trigonometrical consequences. And the trigonometrical inequalities, I generally visualize them geometrically.

MP: You mentioned the Pi Mu Epsilon Journal. *You used to edit the Problems section.*

Bankoff: Sometime in the late sixties, Klamkin was problems editor. But he was bogged down with many other duties. He was editor of the Problems section of the *SIAM Review,* so he asked me to take on the *Pi Mu Epsilon Journal.* I told him I didn't have the ability or the time. A few days later I received a big package from him saying: "You are the editor.

Goodbye." I felt duty-bound to take it over. When I had been editor for a short time, I became aware of the difficulties that have to be faced by problems editors (and you're very familiar with those). I wrote an essay on the problems of a problems editor; it appeared in the Fall 1975 issue of the *Journal*.

One of the earlier editions of Steinhaus's *Mathematical Snapshots* came out in 1950 and contained, among many other things, something that could be called the counterfeit coin problem to ascertain which one of a number of coins is counterfeit and whether it is heavier or lighter. My daughter was a high school student. She looked over this particular solution and discovered an error. I checked it and sure enough there was something wrong with Steinhaus's solution. So I wrote to Oxford University Press and indicated where the error was. In 1960 an edition came out with the error corrected and an acknowledgment. Eight or nine years later another edition came out, and they reinstated the error. It demonstrates what a dentistry professor of mine once said, "There is one entity that has claim to immortality and that is error."

The Man Who Corrected Einstein

There was another curious incident involving an error. It was the spring of 1952. My wife and I were at a movie one evening, and when we came out I picked up a morning paper. There was a big hullabaloo about Einstein and a student who had requested an answer to a geometry problem she had sent him. It made the front page of the *Los Angeles Times:* "Stuck with Geometry, Girl Turns to Einstein: Sophomore Decides Famed Physicist Is Most Apt to Solve Problem—and He Does." Her school was not at all happy with her; the people there thought it looked bad for their teachers if she had to write to Einstein to get help with her geometry problems. There was a picture of Einstein's postcard (and a comment that he didn't put enough stamps on it!). She was pictured with the caption: "Still Puzzled. Johanna Mankiewicz, 15, daughter of Screen Writer Herman Mankiewicz, ponders over sketch supplied to her by Dr. Albert Einstein to help her solve a geometry problem. She said Einstein's sketch didn't help." Well, I looked at the postcard sketch pictured in the newspaper, and Einstein wasn't right either. He gave an answer to a different problem. He described how to construct a tangent to a circle, but the problem was to compute the length of the tangent. All of this was in *The Mirror* as well, so I called the editors and told them the solution given did not solve the problem. They wanted to know my credentials to question Einstein. So they sent a reporter out in the early hours of the morning. The next morning I was on the front page of the *The Mirror:* "Einstein 'Fails' in Math, but Still Has Nobel Prize." I became known around Los Angeles as the man who corrected Einstein!

MP: *Is your interest in mathematics directly related to your interest in dentistry at all? Are you interested in the geometry of the jaw or stress calculations or such?*

Bankoff: No. Only indirectly. I am sort of a Martin Gardner–style debunker. I resent any introduction of the occult into dentistry where so many orthodontists try to attribute occult significance to the appearance of the golden ratio in the human skeleton and dental equipment.

MP: *Are you a subscriber to the* Skeptical Inquirer?

Bankoff: Of course.

MP: *Does that come from your interest in mathematics or your interest in magic or just generally from your outlook on the world?*

Bankoff: Before I knew of the *Skeptical Inquirer,* Bertrand Russell was one of my heroes.

MP: *I know that you are quite close to Sol Golomb and see him regularly. When did you first meet him?*

Bankoff: I met him in 1968 through Leo Moser. I met Leo through Charles Trigg.

MP: *From Golomb in the other direction?*

Bankoff: Oh, it began to fan out in all directions. For example, in the early days when I developed quite a lot of material on the shoemaker's knife, I was called upon to deliver lectures at a number of local colleges—Occidental, Pepperdine, Pomona, UCLA. I got to meet quite a few mathematicians.

One day I got a phone call from Nathan Altshiller-Court, who was visiting Los Angeles. He wanted to meet me, and, of course, I wanted to meet him. He spent a few days with me, and he told me about a forthcoming book of his called *Mathematics in Fun*

Figure 5.9 Bankoff (right) with Howard Eves at Orono, Maine.

and Earnest, which was just about to be published.

One day Leo Moser was in town. He called me and asked me to go over to Caltech with Charles Trigg and him because Erdős was to be there. I buttonholed Erdős, and we took a walk. I had just published my proof of the Erdős-Mordell theorem, which Eves had, I understood later, refereed. One thing led to another.

MP: *What led to your joint paper with Erdős?*

Bankoff: The Putnam Examination in 1967 (see G. L. Alexanderson et al., *The William Lowell Putnam Mathematical Competition/ Problems and Solutions: 1965–1984,* Mathematical Association of America, 1985, p. 7) had a problem that was solved by the use of complex numbers. I received a letter from Donald Coxeter asking me to see if I couldn't solve it using ordinary Euclidean geometry. I was able to do that so I sent it to him and sent a copy to Klamkin. He showed it to Erdős, and we put together this problem with various kinds of solutions, my geometric solution and Klamkin's and Erdős's complex numbers solution. The article appeared in *Mathematics Magazine* in November 1973. I have always been interested in various solutions to a single problem.

MP: This was the asymmetric propeller problem.

Bankoff: Yes. Later we discovered there are quite a few ramifications of this problem. I'm really ashamed of myself for not having completed this. But I did make a special trip to Edmonton to consult with Klamkin about it. He added a few notes about sixfold symmetry that I didn't understand. And because I didn't understand it, I couldn't finish the paper.

MP: Is that one of those papers sitting in your desk drawer?

Bankoff: No. It's sitting in my closet!

MP: Beginning in the forties and fifties you started doing more mathematics. What got shoved aside? Chess? Piano? Photography? (I know you are also interested in photography, enough to own thirteen cameras!)

Bankoff: Piano never got shoved aside. Chess did. In fact, many of these things went on concurrently anyway, but gradually I started spending more and more time on

mathematics. I have at home about twenty-five huge notebooks, just my own notes on problems and solutions and such.

MP: You're working full-time as a dentist, playing the piano, sculpting, playing chess, taking pictures, and doing mathematics. Do you sleep a negative number of hours per night?

Bankoff: No more than four. I understand Erdős sleeps only four hours per night too.

MP: But he pretends to sleep at lectures.

Bankoff: He pretends to, but he doesn't really. He closes his eyes and nods his head, but he's really wide awake.

MP: Do you have any new Erdős stories?

Figure 5.10 Bankoff with Paul Erdős, Beverly Hills, 1980.

Erdős and Bach

Bankoff: The latest Erdős story that I can recount is this. About two years ago, Erdős visited Los Angeles, in the winter, to attend Caltech's Alaoglu Lecture, and, as was his custom, he had dinner one evening with us. Invariably Erdős requests a background of Bach. So we put on some records of Bach, and I happened to mention that the following evening we would be attending a recital at Ambassador Auditorium by Alexis Weissenberg. He would be playing the *Goldberg Variations* among other things. Would he like to attend as our guest? He agreed, and the night of the recital, we picked him up at the Atheneum at Caltech. He was having dinner with Ron Graham. They had papers spread out all over their table, but when we arrived he gathered up his papers, and we proceeded to the Ambassador Auditorium. He made his usual exploratory jaunts upstairs and down to see what was going on there. When the concert started—he was sitting next to me—his head dropped to his chest, and he closed his eyes, as is his custom. To all intents and purposes he went to sleep. After an evening of Bach, we were in a group that went backstage to meet Weissenberg. Erdős wanted to meet him. And, as it turns out, Weissenberg wanted to meet Erdős. They got into a conversation. All of a sudden, Erdős said he had to make a phone call. When he finally got back we asked what took him so long. He said he had had to call Ron Graham. During the concert he had solved the problem they had been working on!

MP: *What was your first publication?*

Bankoff: My first were dental publications when I was still a student. They appeared in a journal, now defunct, called the *Dental Student's Magazine.* One of my pet topics related to occlusion and to changes in occlusal coordination resulting from the loss of teeth. Several articles described unusual clinical findings in my work in the school clinics. After graduation I limited my writing to letters to the editor. Since I neither sought nor desired any academic connections, there was no point in my rehashing the findings of others just to get my name into print. In recent years my extramural activities in dentistry have been political, limited to service on a number of dental society committees.

My first mathematical publication came out around 1950, the solution to a problem.

MP: *You had a nice article about the golden arbelos in* Scripta *in 1955. That's the first full article of yours I am aware of. And you have a nice recent publication in* Mathematics Magazine, *one on the history of the butterfly theorem. What do you have in the works, other than a drawer full of manuscripts?*

Bankoff: Having relinquished my editorship of the Problems section of the *Pi Mu Epsilon Journal*—I turned it over to Clayton Dodge—I have a little more time to spend on unfinished manuscripts that have been lying around.

MP: *So we'll see a number of articles soon. . . .*

Bankoff: I have not only folders of material, but valises of material, large valises. In fact, we're thinking of adding a new structure to our house, to the garage behind the house, to hold all the books, and the manuscripts I hope to finish one day.

MP: *I know that you spoke at the Strens Conference in Calgary. What did you talk about there?*

Bankoff: The arbelos. I called it "The Marvelous Arbelos." It's not that it's the most important thing in geometry, but to me it represents something sentimental. It's the first geometric figure that I devoted great attention to. It is responsible for my trip to Europe and for my meeting my wife.

MP: *Were you reporting on some recent investigations? Or was this a survey?*

Bankoff: Just a survey.

MP: *Shortly after you attended the Strens Conference you attended the International Congress of Mathematicians in Berkeley. And now you are in San Francisco for the joint meetings of the American Mathematical Society and the Mathematical Association of America where you are speaking in a special session on geometry. What are you doing these days?*

Bankoff: Well, not much mathematics. I spend a lot of time with my computers. I have seven, some hopelessly obsolete by now, but I still keep them around. One I use only for playing chess. My daughter, who died in 1985, bought me a Macintosh for Father's Day, just two weeks before she died. So all these years I never even opened the package, and you can understand why. But a few months ago I unpacked it, and I'll have to learn how to use it. I'm still using my IBM 386 clone. I have one of the computer science teachers at Beverly Hills High come in and give me private instruction on the computer, to answer questions that I can't ferret out of books.

MP: *How much of the day do you spend in the office these days? You still have patients who are counting on you.*

Bankoff: I'm trying to cut back. I'm trying not to take on new patients. Of course, people do not really seek out eighty-two-year-old dentists.

MP: *Why not?*

Bankoff: They like to be confident of a sense of continuity!

During the past four years I have taken up jazz piano. I had played classical piano since the age of twelve, but I'm fascinated now with jazz piano. It's such a complex subject. I had studied harmony and counterpoint and felt I understood it pretty well, but I could never understand jazz. It's a different kind of music. It's entirely improvisation based on modern harmonies, voicings, and rhythms—a sort of creative, spontaneous composition. So I take a lesson every Saturday afternoon in jazz piano. I find it quite hard, teaching my hands almost automatically to devise runs, licks, riffs, and left-hand accompaniments.

Chopin's Tomb

MP: *You haven't lost interest in classical music, I suspect. I recall there's a story about your interest in Chopin.*

Bankoff: When I first visited Paris in 1958, I decided that the first thing I wanted to do was visit Père Lachaise Cemetery to see Chopin's tomb. I was not in the best shape. I was recovering from a broken leg and had been on crutches till very recently. So I called a cab and asked the driver to take me into the cemetery, but that was not permitted, at least

until I reached into my pocket for a few bills, and suddenly things become possible. I got some flowers to take in. It was the only grave in that area to be covered with fresh flowers. I found that the muse on the tomb, with head bowed, was beautiful except for one thing. Somehow—maybe it was the work of a vandal—one finger had been broken off her hand. So when I got back to Los Angeles, I asked a patient of mine, the novelist Romain Gary, who was French consul-general there at that time, what I could do to repair the statue. He got me in touch with the estate of Chopin, and they replied that they could find a sculptor to do the job. Actually, maybe they misunderstood my question. They also gave me an estimate on what it would take to restore the whole tomb. That was not my

Figure 5.11 Bankoff placing flowers at Chopin's tomb, Père Lachaise, Paris, 1968.

intent. So I sent them a check to have the finger repaired. I never heard from them, but a year later I was in France again and went out to Père Lachaise. I was dismayed to find that I could have done a much better job myself. Instead of having a nice graceful curve to the finger, it was sticking straight up in the air. I almost felt like breaking it off myself. But I didn't. A couple of years later I returned to find that someone had broken off the offending finger and the thumb as well. That's it. I'm not repairing it again!

MP: *Well, you did what you could.*

So at this point in your career, do you have any regrets? Did you choose the right career?

Bankoff: My only regret is that I have so little time to spend on mathematics, the most beautiful thing in the world.

MP: *But choosing dentistry as number one and mathematics as number two—is there any reason to believe that you should have chosen them the other way around?*

Bankoff: Well, I am a mathematician without being a professional mathematician. I don't look upon myself as an amateur. Someone defined an amateur as one who does not

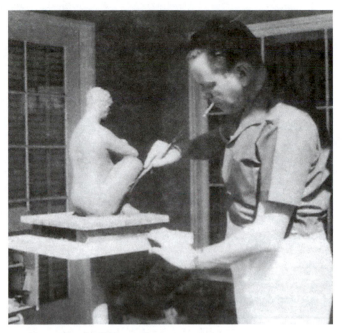

Figure 5.12 Bankoff, the sculptor, 1955.

earn money with what he is doing. But it is an activity that I indulge in as thoroughly as other mathematicians.

MP: *But I like to associate "amateur" with its root, to mean that one loves the subject.*

Bankoff: I certainly do. Of all of my interests, I put mathematics number one. How I make my living is another story.

Six

Alice Beckenbach

Alice Curtiss Beckenbach came from a mathematical family. She was born in 1917, two years after the Mathematical Association of America (MAA) was founded, and over the next eighty-five years she was a regular attendee at winter and summer national MAA meetings. Alice's mother Sigrid Eckman was president of her class at Radcliffe, a class that included Helen Keller. There she met Alice's father David Raymond Curtiss, who completed his PhD in mathematics at Harvard in 1903. Curtiss spent his academic career at Northwestern University, many of those years as department chairman. He was author of the second Carus Monograph (on complex analysis) and was president of the MAA from 1935 to 1936. David Curtiss's brother Ralph was an eminent astronomer at the University of Michigan.

Alice's brother John Curtiss, who had a PhD in mathematics from Harvard, headed the Applied Mathematics Division of the National Bureau of Standards from 1946 to 1953 when the Division was charged with developing mathematical methods for exploiting the first generation of electronic computers. He was the founding president of the Association for Computing Machinery and later served as executive director of the American Mathematical Society.

Alice was a dual mathematics-chemistry major at Northwestern and about to start graduate study in mathematics when she met her first husband, Albert W. Tucker, who was teaching summer school at Northwestern. Tucker was a well-known topologist at Princeton who later turned to mathematical programming and game theory, famous for Kuhn-Tucker theorem and the prisoner's dilemma. Tucker was chair of the Princeton mathematics department in the 1950s and 1960s. He instigated the now-famous common room teas at Princeton. His PhD students and postdocs included two Nobel laureates in economics, John Nash and Robert Aumann, along with Marvin Minsky, the founder of the AI Laboratory at MIT. Al Tucker was president of the MAA from 1961 to 1963 and was one of the early recipients of the MAA Distinguished Service Award.

Alice's two sons, Alan and Tom, are mathematicians and have served the MAA in many roles; both have been MAA first vice-presidents. Her third child, Dr. Barbara Cervone, is a prominent educator who currently heads the nonprofit group What Kids Can Do. Alice has two grandchildren with PhDs in the mathematical sciences: Thomas J. Tucker, a number *theorist at the University of Rochester, and Lisa Tucker Kellogg, a computational biologist at the National University of Singapore.*

In 1960 Alice married Ed Beckenbach, who spent most of his career at UCLA, where he was chair of the mathematics department and a founder of the Pacific Journal of Mathematics. He also was a co-author of the Dolciani K–12 series of "new math" textbooks, widely used in the 1960s and 1970s. Ed was the chair of the MAA Publications Committee for over a decade. Posthumously, he was also a recipient of the MAA Distinguished Service Award.

Alice is now 92 and no longer able to attend MAA meetings, but her heart and soul will be with mathematicians whenever they gather. [She died on March 18, 2010.]

Alan C. Tucker
Thomas W. Tucker
September 2009

Reminiscences of a Mathematologist

A mathematologist may be defined as someone who specializes in the study of mathematicians and their peculiarities.

It is no secret that the mathematics professor is a breed of human both admired and shunned by the ordinary person ("and perdaughter," to quote Ralph Boas) and locked lifelong in a love-hate relationship with the remainder of the human race. The "victim" body count in my own family thus far includes my father, brother, two husbands, two sons, and now at least three grandchildren on their way to joining the list. Perhaps I should add that although treatment for my family disease of mathematics is mainly supportive, there is no known cure for this

Figure 6.1 Alice with sons (L to R) Tom and Alan and Andy Gleason.

puzzling abnormality other than the radical surgery of amputating the entire head. Needless to add, medical and legal experts frown on this crude type of instant therapy.

Accordingly the voice of experience may be permitted to share some reminiscences from a lifetime of close acquaintance with the world of mathematicians. Many of their personality profiles will be punctuated with examples of verse from the unpublished works of several of their colleagues, who would doubtless prefer to remain anonymous.

To the average layman, an obvious peculiarity of the serious mathematician is a frequently glassy-eyed expression along with a tendency to absentmindedness owing to intense preoccupation with inner thought. Probably the most famous example of the absent-minded mathematician was that lovable old sandal-clad hippie Albert Einstein, who could never remember such finishing sartorial touches as combing his hair or donning a pair of socks. When he took up residence at the Institute for Advanced Study in Princeton, the university students used to sing a quatrain in his honor with these words:

> *Here's to Einstein, Albert E.,*
> *Father of Relativity.*
> *You'll know him by his fiddler's locks*
> *And by his utter lack of socks.*

Thus it was that one snowy morning some years ago when my then nine-year-old son, Tom Tucker, the mathematician-to-be, was leaving for school, I was proud to observe the glassy-eyed look on his face as he waved good-bye and headed out into the wintry blast. The fact that *he* had on *socks* as he trudged through the snow, although *no shoes,* made me doubly proud that *our* budding Einstein at least had some common sense. So, you see, I too at least had symptoms of the family disease.

Another example of such absent-mindedness is the story about one of the greatest mathematicians of this century, David Hilbert, who was a professor at Göttingen University, the leading shrine of mathematics in the world in the first quarter of the century, until Hitler

Figure 6.2 Einstein with Adolf Hurwitz and daughter Lisi Hurwitz.

Figure 6.3 Who needs shoes? Tom Tucker at age nine.

drove out the most brilliant minds there. In those days, a new member of the faculty there was supposed to introduce himself to his colleagues in a very formal manner: he put on a black coat and a top hat and took a taxi to make the rounds of the faculty houses. If the Herr Professor was at home, the new colleague was supposed to go in and chat for a few minutes. Once such a new colleague came to Hilbert's house and Hilbert decided (or perhaps his wife did it for him) that he was at home. So the new professor came in, sat down, put his top hat on the floor, and started talking. . . . All very well, but he didn't know how to stop talking,

and Hilbert became more and more impatient, eager to get back to an important theorem he was inventing that day. Finally, Hilbert stood up, picked up the top hat from the floor, put it on his own head, touched the arm of his wife and said, "I think, my dear, we have delayed the Herr Professor long enough"—and walked out of his own house.

> *Joseph Wedderburn, so they say,*
> *Has Einstein beat in every way.*
> *He sublimates the urge of sex*
> *By playing with the powers of x.*

Obviously, Professor Wedderburn would not have qualified to be included in the *World Almanac* table on labor statistics under the heading: "Number of People in the U.S. Gainfully Employed by Sex."

My father, D. R. Curtiss, was a very active member of both the MAA and American Mathematical Society in the first half of the [twentieth] century. In the mid-1930s, when

Figure 6.4 Hilbert—not always patient.

Figure 6.5 D. R. Curtiss, father of Alice Beckenbach.

Figure 6.6 Ping pong buffs—Garrett Birkhoff (left) and Hassler Whitney (right).   Figure 6.7 Lars Ahlfors.

he was president of the MAA and I was studying mathematics in college, I often accompanied him to the winter and summer joint mathematics meetings. There my main contribution was in the practice, not the theory, of games, offering a variety of amusing diversions to some of the more playful attendees. I fondly recall, for example, regular ping-pong challenges with Warren Weaver at AAAS meetings, tennis with Nathan Jacobson, Garrett Birkhoff, Hassler Whitney, and Marston Morse (among other buffs of the game), and also an occasional piano duet with the latter, and bridge games with various others.

At the 1935 winter meeting in St. Louis, Professor Lars Ahlfors, first winner of the Fields medal, had just arrived from his native Finland to join the faculty at Harvard. He requested and I gave him his first driving lessons on a

narrow, icy road along the Mississippi River. So clearly, this "math groupie" had courage, too, along with her more playful talents.

Another of my recollections of the meetings in those days was the fearsome triumvirate of Lefschetz, Tamarkin, and Wiener sitting in the front row at the ten-minute talks given by young new PhDs who were locked in a desperate scramble for teaching jobs during the Great Depression. Lefschetz would be as usual alert and on the edge of his seat, Wiener blissfully snoring, and Tamarkin poised somewhere in between, all three waiting to pounce on the hapless victim with such ego-busting remarks as, "After I obtained that result back in—think it was 1926—I found some interesting extensions that I'd like to mention. . . . "

In 1938 my career as a mathematologist soared to heady new heights when I left my

Figure 6.8 Wiener, one of the Gang of Three, who could constitute a tough audience.

Figure 6.9 Luther P. Eisenhart served as Department Head, Dean of Faculty, and Dean of the Graduate School at Princeton.

childhood town of Evanston, Illinois, with its then isolationist, ultraconservative, church-on-every-corner, national-headquarters-of-the-Woman's-Christian-Temperance-Union mystique, to become the bride of A. W. Tucker, a mathematics professor at Princeton University. The town of Princeton was small but cosmopolitan and something of a Grand Central Station for mathematical refugees from Nazi Europe. My husband's office was sandwiched between Einstein's on one side and that of the noted physicist Niels Bohr on the other. Among the resident notables whom one often spotted strolling up Nassau Street at the time were the writer Thomas Mann, the pianist Robert Casadesus, and, a few years later, Bertrand Russell.

The Institute for Advanced Study was well established then and shared quarters with the Princeton mathematics department in Fine Hall—and a fine hall it was, with a couch and

fireplace in every office, plus a men's room with full bath. The department chairman was Luther P. Eisenhart, who was honored with the following quatrain:

> *Here's to Eisenhart, Luther Pfahler,*
> *In four dimensions he's a whaler.*
> *He built a country club for Math*
> *Where you can even take a bath.*

The tenured mathematics faculty consisted of Eisenhart, S. Bochner, H. Bohnenblust, C. Chevalley, A. Church, S. Lefschetz, H. P. Robertson, A. W. Tucker, J.H.M. Wedderburn, E. Wigner, and S. S. Wilks. Instructors or postdoctoral fellows were N. Steenrod, Ralph Fox, J. Tukey, Brockway McMillan, and C. B. Tompkins, among others.

At tea time the Fine Hall common room was a gathering place for the mathematicians and mathematical physicists (probably and

Figure 6.10 Did Paul Erdős live in the Fine Hall Common Room?

Figure 6.11 Lefschetz—The Great White Father.

Figure 6.12 Albert W. Tucker

most notable among the latter being Richard Feynman, then a graduate student), and a center for spirited games of kriegspiel and arguments about Bourbaki. It was also apparently the living quarters of Paul Erdős; at least, I remember opening a drawer in one of the cupboards there and coming upon his underwear and pajamas casually stashed away. Alas, the life of a proper mathematics department chairman is not an easy one, and it particularly rankled the dignified Professor Eisenhart to see Erdős gleefully sliding down the banisters in Fine Hall whenever he had just "sunk up a new serum [thought up a new theorem]."

Lefschetz, too, kept the air crackling around Fine Hall, needling his adversaries and amusing his friends with his quick wit and outspoken opinions. The Great White Father, as he was fondly known to his graduate students, was memorialized in verse as follows:

Papa Lefschetz, Solomon L.,
Irrepressible as hell,
When laid at last beneath the sod,
He'll then begin to heckle God.

My then husband, Albert Tucker, was a protégé of Lefschetz and the appointed peacemaker between the math department and the university administration when Lefschetz succeeded Eisenhart as Chairman. The engineering department, in particular, detested the math department, which in turn had naught but contempt for the engineers. Bitter indeed were their battles over the teaching of calculus. Now, Tucker was a careful, responsible sort of person, and a notable exception to the prototype victim of mathematicitis. He was not absent-minded, and he could count beautifully. In the following quatrain describing his trials and tribulations, the name Pétard refers to a group of six young Princeton Turks who submitted to the *Monthly* an anonymous article that was a clever spoof on Bourbaki. But the *Monthly* editor was loath to accept anonymous articles, and something of a brouhaha ensued. The article was then submitted under the name H. Pétard. So, on with the verse:

Tucker likes to keep things straight,
Topologize, and not be late.
But in these things his life's made hard
By Alice, Lefschetz, and Pétard.

Permanent mathematical members of the Institute for Advanced Studies in the late 1930s were, in addition to Einstein: J. W. Alexander, Kurt Gödel, Marston Morse, John von Neumann, Hermann Weyl (whose poetic contribution concludes with:

Figure 6.13 Hermann Weyl in a not too saintly pose.

Figure 6.14 "Uncle" Oswald Veblen gave rather formal dinner parties.

For he is that most saintly German,
The One, the Great, the Holy Hermann.)

and not least of all, the magnetic mathematician-promoter and stately but genial anglophile, Oswald Veblen. Married to the sister of a famous British physicist, he, too, was immortalized in doggerel to this effect:

Hail to Uncle Oswald V.,
Lover of England and her tea.
He's the only mathematician of note
Who needs four buttons to fasten his coat.

"Jimmy" Alexander was in many respects the Renaissance man of this notable group. He was a charming and ruggedly handsome millionaire, a skilled mountain climber, figure skater, ham radio expert, music lover, limerick collector, and a liberal who even

Figure 6.15 James W. Alexander giving a ride to a delighted Lefschetz.

marched in the Communist May Day Parade in New York.

Social entertaining at that time was usually joint between the mathematics departments of the University and the Institute. A dinner party at the Veblens was a highly formal affair, with enough silver at each place setting to summon the Lone Ranger's horse. After dinner there was demitasse for the ladies in one drawing room, while the gentlemen amused themselves over brandy and cigars in another.

Entertainment at the von Neumanns', by contrast, was far more informal. For example, the wives were included when the men gathered for some lively after-dinner raconteur competition. The von Neumanns' preference for casual entertaining hit something of a peak at one of their wartime parties when they recruited Army Specialized Training Program servicemen wandering by on the street to join in the carousing inside. Vivid in memory, too, was a light-hearted party there in honor of the von Neumanns' daughter Marina, now a distinguished economist, on the occasion of her fourteenth birthday when she had just come to live with him and her stepmother, Klara. A book recently published, entitled *The Prisoner's Dilemma* by William Poundstone, gives sort of a biography of John von Neumann, as well as of game theory.

Another von Neumann incident that I recall with some amusement occurred at a dinner at the home of Professor Emil Artin and his wife Natasha. The guests were seated at card tables. At my table there were just three of us: two mathematical greats, von Neumann and Weyl, and one mathematologist. Norbert Wiener's pioneering book *Cybernetics* had just been published, and von Neumann had been asked to review it for a leading journal.

Figure 6.16 John von Neumann gave somewhat informal dinner parties.

Weyl asked von Neumann for his opinion of the work, so the latter commented in true Hungarian style on its applications and implications, or rather on a lack of any, according to him. "But you know," he added, "Wiener has zees inferiority complex, so when I wrote zees review . . . well, to tell zee truth, I just couldn't!" It was von Neumann who claimed to have invented the popular definition of a Hungarian, namely: "It takes a Hungarian to go into a revolving door behind you and come out first."

Artin was another eminent mathematical refugee from Germany. His son Michael was President of the American Mathematical Society (1991–1992) and a well-known mathematician at MIT.

Stories about Johnny von Neumann are legion, but unfortunately no one seems able to recall any that were put to verse. One of his own favorite verses was this:

Figure 6.17 Marston Morse—a confident man.

There was a young man who cried "Run!"
The end of the world has begun!
The one I fear most
Is that damned Holy Ghost
I can handle the Father and Son.

Certainly he was an amusing raconteur and a notoriously wild driver, with gaping holes in his neighbor's hedge to prove it. Perhaps that sheds some light on this line from a description of Princeton once given by the noted Institute physicist Freeman Dyson: ". . . where every prospect pleases, and only man is vile." Nevertheless, it was generally agreed that von Neumann was the leading mathematical mind in the world at that time.

Then there was Marston Morse and his charming wife Louise, who were prime hosts to both resident and visiting mathematicians. Dinner at their home was always a delightful and delicious experience. The dynamic, multitalented, aquiline-nosed Morse, author of the definitive text, *Calculus of Variations in the Large,* was not known for a lack of self-confidence. For example, once at a dinner party he was seated next to the wife of a Princeton faculty member. She was a matter-of-fact, no-nonsense sort of person. A bit of polite exchange revealed the fact that they had grown up in the same town in Maine and that both had relatives with the same name. This prompted Professor Morse to comment with a touch of courteous enthusiasm, "You know, we might very well be cousins!"

"Oh, that's ridiculous," she replied firmly. "I doubt very much that we're cousins."

Shortly after, as the guests adjourned to the living room, Morse took his wife aside and whispered, "Say, what does that woman *want* in a cousin?"

Thus it is that the rather wicked quatrain in his honor ends with the lines:

His opinion of himself, we charge,
Like nose and book, are in the large.

But enough of this careening down memory lane. It's still not too late to heed the sage advice that my father once gave me: "It's simply not true that you have to say *everything* that comes into your head."

Seven

Arthur Benjamin

Donald J. Albers

Arthur Benjamin is a Professor of Mathematics at Harvey Mudd College in Claremont, California. He is a gifted teacher, an accomplished magician, and a dazzlingly good mental calculator. Now forty-eight, his career has been marked with a string of successes. But as a child growing up in Cleveland, Ohio, all bets were off about his future. Little Art Benjamin was so rambunctious in his first nursery school that he was thrown out! Ditto for his second nursery school, and the third. Today his problem would be called ADHD. He overcame the problem, eventually earning a PhD in mathematical sciences from The Johns Hopkins University.

Benjamin's passion for both magic and mathematics is so strong that he refers to himself as a "mathemagician." As a teacher, he has a knack for involving students with his subject, and he has written many research papers in collaboration with them. During his spare time, he performs on numerous stages as Art Benjamin—Mathemagician. He has appeared on television shows in the United States, England, Canada, and Japan, including the *Today Show, Evening Magazine, Square One,* and on CNN. He has been profiled in the *New York Times,* the *Los Angeles Times, USA Today, Scientific American, Discover, Esquire, People,* and several other publications. In 2005 *Reader's Digest* called him "America's Best Math Whiz."

Benjamin has a strong interest in the theater, which is not surprising in view of the fact that his brother is an actor and director, his sister has a trained voice, and his father was an amateur actor and director as well.

His performing abilities and awareness of audience involvement serve him well in the classroom. It is common for him to analyze one of his class presentations to see how he can increase student participation. He has won significant teaching awards, including the Haimo Award for Distinguished College or University Teaching from the Mathematical Association of America (MAA), in 2000. After receiving his doctorate, he sought a school that would value his teaching as well as his research. He says that he found it in Harvey Mudd College.

Benjamin has written more than seventy research papers, most in combinatorics, game theory, or number theory, and two books: *Secrets of Mental Math,* with Michael Shermer (Three Rivers Press, 2006), and *Proofs That*

Really Count: The Art of Combinatorial Proof, with Jennifer Quinn (MAA, 2003), which won the Beckenbach Book Prize from the MAA. In 2009 he co-authored the MAA book *Biscuits of Number Theory* with Ezra Brown.

On the occasion of the 1998 Mathfest of the MAA in Toronto, Arthur Benjamin, together with Brent Morris, gave a special short course on magic in mathematics and, following that, at the opening banquet, he performed feats of mental calculation, to sustained and enthusiastic applause. This interview took place the following day, July 17, 1998.

Childhood

MP: *You were born on March 19, 1961, in Mayfield Heights, a suburb of Cleveland, Ohio. You have a brother and a sister. What did your parents do?*

Benjamin: My father, Larry, was an accountant, and my mother, Lenore, was a special education teacher. My hyperactivity as a child drove her to get a master's degree in special ed.

MP: *Really?*

Benjamin: Yes, so she could understand me.

MP: *Let's start out, then, with Arthur the child.*

Benjamin: As a child I was hyperactive, and I think in today's language, they would say such a person has ADHD, attention deficit hyperactivity disorder. I love attention, and I was a tough kid to handle. I was kicked out of several nursery schools.

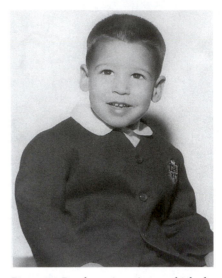

Figure 7.1 Rambunctious Art was kicked out of his first nursery school.

MP: *You were literally kicked out?*

Benjamin: Yes, just because it would be nap time, and all the kids would dutifully lie on their blankets, and I would be running around the room. I think looking back, a lot of the problem may have stemmed from the fact that I was bored with many things.

I understood things the first time they were explained, and by the time something was taught or done the fourth or fifth time, I'd rather get silly than pay attention. There was a lot of that, but there was also something medical about it. Back then, the treatment was to put you on Valium, which I took for ten years, until I was in eighth grade.

MP: *But it did calm you down.*

Benjamin: Yes, I think so. My parents thought there was a big difference, but the dosage was being reduced gradually. I don't know to what extent the effects of the hyperactivity are still there. My wife says I'm certainly easily distracted and to some extent have a short attention span. People find it interesting when they see one of my shows and are surprised to know that I am very absent-minded. I was late for this interview, partly because I got lost. What I'm pretty good at is concentrating in bursts, which makes games like chess and backgammon good games for me and allows me to mentally multiply five-digit numbers. That kind of thing I can do, but sustained attention is another matter. For instance, if that television were on, even with the sound off, I'd be constantly looking over there.

MP: *You must be a serious channel surfer then.*

Benjamin: Well, no. It irritates me when people channel surf. My preference is to keep the TV off, but I tend to get sucked in. As a child, I wanted a lot of attention, and I would do lots of things to get it.

MP: *What, for example?*

Benjamin: Finding magic as a hobby provided something to show off. I would sing. I tried dancing. I became very good at different games. I learned to calculate quickly— things that would impress people. I used to memorize things. When I was five years old, I memorized the states in alphabetical order and their capitals. I learned the presidents in order. Later in life, I learned the Tom Lehrer songs and memorized the elements.

MP: *Clearly your parents were giving you lots of attention. After all you were their first-born. The first child usually gets more attention.*

Benjamin: Actually, in my first four or five years of life, I spent a lot of time in children's hospitals, to diagnose and treat my hyperactivity.

MP: *For what length of time? Days at a time? Weeks?*

Benjamin: Weeks.

MP: *So you were seriously hyperactive.*

Benjamin: Yes. I can remember being in one hospital that was especially traumatic. They put a net over my bed because I would crawl out at night and walk down the halls. I hated that.

MP: *You felt as if you were trapped.*

Benjamin: Who would want to sleep in a cage, which essentially it was. But for all my hyperactivity, there was no maliciousness.

MP: *You weren't destructive.*

Benjamin: No, I was a nice gentle kid, probably just overly curious.

MP: *Did you have the other usual interests in terms of playing in sports?*

Benjamin: Oh, yes. I liked sports. I wasn't very good because the medication did slow me down. I do think that it made me less coordinated. When I got off the medication, I went out for the track team. My parents

Figure 7.2 Mr. Benjamin, dressed for success.

exposed me to a lot of different activities and let me go in the directions I had aptitude for and enjoyed the most. There was no push to make me a mathematician or my brother Stephen an actor or whatever.

MP: *Is he an actor?*

Benjamin: He was for a time. Now he's a teacher. But what the three of us have in common is a profound love for the theater and being on stage generally. In fact, last night my wife Deena and I were at the Shaw Festival. We went down to Niagara on the Lake and saw *Major Barbara*. It was wonderful. From the time we were about eight years old or more, my siblings and I were on stage, performing in the community theater.

MP: *Did your mother and father also perform?*

The Great Benjamini

Benjamin: My father was very active in community theater. He was an accountant by day and an actor and director by night. I think he wished that he had taken a shot at some form of career in theater, either as an actor or as a theatrical manager. I think he passed that on to us. I think that's why my brother tried for many years to make it in the entertainment industry and gave it a very good shot. He's very, very talented, but he came back home to Cleveland where he's still one of the stars of the community theater scene as an actor and director. He followed somewhat in Mom's footsteps and got his master's in special education.

My sister Mara does public relations advertising, and she also is still very active in theater in Columbus, Ohio. She has a trained voice. My mother never did anything on the stage, but she was very supportive of all of us in that regard. From a very early age we had not only no fear but an enjoyment of being in front of groups of people, and that has helped me immensely in my teaching.

I did my magic as a hobby in high school. I did magic shows throughout the East Side of Cleveland. My stage name was The Great Benjamini.

MP: *The Great Benjamini! Was this your father's suggestion?*

Benjamin: I don't know who suggested the name. He printed a business card for me once. I think we had used it as sort of a joke. I actually did shows, birthday parties, etc. I didn't have much in the way of expensive

Figure 7.3 As a high school student, "The Great Benjamini" performed magic.

equipment. I just did what I could to make the kids laugh—fall on my face, and slapstick. That's what the kids like, and I just did things that involved a lot of audience participation. I wasn't just up there doing a trick. The real trick was getting the audience involved.

MP: *You're still doing that.*

Benjamin: Yes, that's what I'm still doing as a teacher.

MP: *And that really works.*

Benjamin: Yes, and as a performer, too. You don't want people to hear just monologues. Those are rarely great. Dialogues, conversations, active involvement, hands-on—at least minds-on—involvement, and whether I'm teaching or performing, I like to do that. So I think I learned a lot about teaching by entertaining six-year-olds.

MP: *From kids?*

Benjamin: Yes, because if you don't involve them, they'll tune out. So when I started getting some teaching experience, while I was an undergraduate at Carnegie Mellon, I tried to involve the students.

MP: *As an undergraduate?*

Benjamin: Yeah. I was a teaching assistant (TA) for an economics class, for a professor who felt that the undergraduate students from his class could do a better job as TA than graduate students who hadn't had his class. I also did some TA work for the statistics department, which was my area of concentration.

MP: *Was that your undergraduate major?*

Benjamin: Yes, I got a BS in applied math, concentrating on statistics at Carnegie Mellon. I did not have that much pure mathematics exposure. I had to learn more pure mathematics when I started graduate school in operations research at Cornell. I consider myself on the border between pure and applied math. Discrete mathematics is my love; both applications and theory are very attractive to me. The types of courses I teach at Harvey Mudd are calculus, discrete math, linear algebra, abstract algebra, number theory, operations research, and game

Figure 7.4 Benjamin appeared on the Colbert Report in January 2010.

theory. It's a broad spectrum, not overly pure and not classically applied.

MP: *I want to get back again to your early involvement with magic. You said that kids appreciated seeing you do rather standard tricks. And you became very sensitive to what works with kids and what doesn't.*

Benjamin: When I started doing shows, I didn't have much in the way of fancy equipment so I was really entertaining them with myself, my own personality, just monkey business, having fun, making intentional mistakes. Kids love watching that sort of thing.

MP: *The other night after your performance at the opening banquet several people came up to talk with you, but those who seemed to hang around the longest were kids. They seemed to be intensely interested in what you were doing. Let's trace the origins of your interest in magic.*

Early Interest in Magic

Benjamin: As a kid, I think I did lots of things just to show off. I think by about seventh or eighth grade, I was still doing magic as one of my hobbies, and nobody else in my school was. So that did get me a lot of attention.

That gave me an incentive. I realized that I was somewhat special in that regard. I really worked hard to do more of it. The exact same things happened with my mental calculations training. I realized that I could do this, and we weren't going to learn this in school. I guess I just had a personality that worked well with magic, whether it was for six-year-olds or my peers. I would learn various card tricks and coin tricks, and people were impressed. That was a good thing for a couple of years. In fact, it really helped socialize me. I was a studious little kid.

Doing magic tricks became such a social boom for me that it eventually became a crutch, to the extent that I wouldn't go anywhere without a deck of cards in my pocket. I wouldn't go to the corner drugstore without something in my pocket, just in case the pharmacist wanted to see a trick.

MP: *Wow!*

Benjamin: I think it started as something like, "Hey, this guy is interesting." Then it became "Make him stop!" but I was somewhat oblivious to that. I'd think, "Hmm, I'm not impressing people the way I used to. Maybe I'll have to learn more or get better at this or that." As a consequence, I learned a lot of interesting skills and impressive talents, but all the while, maybe I was hurting myself socially. It was sort of "enough already, enough." I don't try so hard to impress anymore.

And that's why it was good when I went to college, Carnegie Mellon, where nobody knew me. I got off to a fresh start. I didn't want people to know me for my magic.

MP: *You concealed your magic?*

Benjamin: Pretty much.

MP: *Can you recall why you made that decision before going to Carnegie Mellon?*

Benjamin: I wanted to make friends, and I wanted a fresh start. But in the spring of my first year of college, there was a magicians' convention in town. That wasn't unusual, but the headliner was Harry Lorayne, whom I really wanted to meet. Lorayne was an outstanding magician and had written some of the best books on card magic ever. I dutifully studied from them in high school. He was also famous as a memory expert. He had written a couple of books on how to improve your memory, including one called *The Memory Book,* that he co-wrote with Jerry Lucas, which was a best seller.

My mother in her special education studies showed me the book, and I absorbed everything in it. I still incorporate some of these techniques when I'm doing large mental calculations. You have heard me turning numbers into words; that's using a mnemonic code system that I learned from his book. It's called the major system, and it has been in use in the English language since at least the 1870s.

As a high school student, I taught a one-week course on how to improve your memory using the techniques that I learned from Lorayne's book. Apparently teaching was something I always enjoyed doing.

MP: *Teaching for me is performing to some extent.*

Benjamin: Yes. By then, of course, I had been in front of an audience for a long time. When I really started teaching, I had the same expectations that a performer would have. You want to keep your audience alive. You even want to keep them laughing to some extent, if you want them engaged in what you're doing.

You don't want them falling asleep on you (unless you are a hypnotist). So I worked hard at the teaching to maintain the same kind of enthusiasm.

MP: *It's very important to you?*

Benjamin: Absolutely! If I come out of a lecture, I can tell when it fell flat. I will go back and analyze the lecture and think of what would be a better way to do it, because if they don't take things away from your lecture, you can be replaced by a well-written

Figure 7.5 Mathemagician Benjamin continues to perform amazing feats of mental calculation on television and in many other settings.

set of notes. What's the point of being up there if students are not going to get something out of classes? Incidentally, students are rational people. Generally, they know they don't have to come to class. If you're not going to give them good use of their time, then they won't show up.

By the way, just on my way over here, I finished reading the MAA book, *Lion Hunting and Other Mathematical Pursuits* by Ralph Boas, and his chapter on teaching resonated with me fully. Boas was a guy who understood teaching!

MP: *Let's get back to the magic convention. You got a lot of attention there.*

Benjamin: Yes. I demonstrated some of my mental calculations for some of the magicians at this convention, and the reaction was bigger than I expected. In particular, there were some guys who were opening up a magic nightclub in downtown Pittsburgh called the Dove and Rabbit, which has long since closed. They asked if I would be willing to perform. And I thought, why not? It was separate from my college life. So nobody knew that I would take buses to downtown to perform.

MP: *You were really keeping this under wraps then?*

Benjamin: Yes. It was in the spring of my freshman year when I performed there every two or three weeks and would do two or three shows a night. While there, I really developed my stage routine. It's the same show that I perform for general audiences now at the Magic Castle in Los Angeles. I met other magicians who also were performing there. They got me involved in the Pittsburgh magic community, and they were an extremely supportive group of magicians, including another famous magician named Paul Gertner. He took me under his wing. He said, "You've got something special here. You've got something that other magicians work all their lives trying to get." Then I attended other national magic conventions in Evansville, Indiana, and Colon, Michigan. Pittsburgh was hosting the International Brotherhood of Magicians meeting the following year, and I was one of the opening-night performers. That got me a lot of exposure in the magic community.

Magic or Mathematics?

MP: *How big is the magic community?*

Benjamin: The national meeting of the IBM (International Brotherhood of Magicians) had a thousand or so magicians in attendance.

MP: *That's a lot!*

Benjamin: My act got written up in some magic magazines. I was in college, and I was even giving thought to becoming a professional magician.

MP: *And giving up mathematics?*

Benjamin: I hadn't thought seriously of dropping out of college, but I did think about magic as a career when I graduated. Anyway, the Dove and Rabbit gave me a chance to repeat the same kind of show and refine it. That was good. The Pittsburgh magicians, notably Paul Gertner, were very supportive. I did not get that kind of support in Cleveland. I performed at the Pittsburgh magicians meeting the following year. I was introduced by a guy named James Randi. And that was a turning point for me. He's an amazing magician and known as a challenger of paranormal claims—he has been called Psychic Enemy No. 1. He won the McArthur Award in 1986.

He made a rare appearance at this Pittsburgh meeting and somebody said, "Hey, Randi, have you ever seen Benjamin? You've got to see him." So, I go and show him my stuff, and his first instinct, of course, was to ask, "How is he doing this?" There are a number of ways that this could possibly be duplicated. I could be using a fake calculator, which has certain answers preprogrammed. I could have an assistant somewhere, who is communicating the answer to me. The first thing he did was to give me a few problems that combine magic and logical thinking. I was able to answer them, which impressed him. Then I started explaining my methods to him, and he concluded that I was for real. And he got very excited because his reputation had been one of a somewhat negative person. People thought of him as someone who went around saying, "you can't, you can't, you can't," and here's somebody who can. This is the sort of phenomenon that we should be focusing our attention on, not people who claim to do things that break all the rules of math and science. These are the positive things that we should be giving our attention to, not astrology and psychics and all that stuff.

So he took me on the road. He said, "I've got people I want you to meet. I want to introduce you to Scot Morris of *Omni* magazine. I'm friends with Leon Jaroff who is the founding managing editor of *Discover* magazine." The *Discover* article about me was a direct result of Randi bringing me up to Jaroff's office. That's Randi. He got me a show at a public library just so that there could be a photographer there to take pictures, and the funny thing is that this picture has made its way into a mainstream textbook on cognitive psychology. So that got me a certain amount of attention. I was mentioned in *Omni* magazine as well, and that was where the early exposure came from. Randi and I have remained close friends. He has been very supportive. He took me with him to Japan in 1989, where we did a television special together.

Between Randi and the magicians in Pittsburgh, I gained new confidence in my magic, but the third part of the equation came in my freshman year of college. I was taking a course in cognitive psychology, which I was always interested in. The whole business about how we think and memorize is fascinating. I was very excited to be taking that course. The professor, Marcel Just, was lecturing on human calculators one day

and what common properties they had and what these individuals could do. He himself had learned a few tricks that are designed to look as if you have mental talent. He said, "Now, I know we have a lot of science and math students and engineering students in this classroom. Does anybody here know any tricks?"

Mental Calculation

MP: *You just happened to have a few tricks to demonstrate.*

Benjamin: I really shouldn't have done it. There I was trying to keep things under wraps, but, when somebody said that, I had to say yes. So I got up and took over the class for the next ten minutes doing my magic act from the Dove and Rabbit. That made a big impression on the class, and on the professor. Afterwards, the professor told me that as part of most undergraduate psychology classes, students are required to participate in a handful of experiments for their professors and graduate students, who are doing research.

I asked him whether there were any professors who would like to talk to me about my mental calculating instead of just doing some random kinds of experiments. He said, "I think I know someone who would be interested." So he introduced me to Dr. William G. Chase, who was an expert in skilled memory and was working with a student who had expanded his memory considerably. The student had expanded his recall of digits from groups of eight digits up to

Figure 7.6 Benjamin outperforms many calculators—even very big calculators.

seventy digits. Dr. Chase was interested in my use of mnemonics and other techniques. I did a few sessions with him. I became something of a research assistant and research subject for the next few years. In the process I acquired a better understanding about what I was doing and my capabilities and my limitations.

When I started college, I did not think that multiplying distinct three-digit numbers was something I could ever do in my head. I could square them, but multiplying different ones seemed beyond me. By the time I graduated from college, I was multiplying different six-digit numbers together without having to see the problem.

The problem would have to be called out to me slowly so that I could create mnemonics to remember the twelve digits of the problem, and somewhat laboriously I would be able to get the answer correct over half of the time. Nowadays in my finale, I will either square or multiply two five-digit numbers.

MP: *But you can do six?*

Benjamin: I can, but I've never publicly performed them because the error probability is too high. Because the college was interested in what I was doing, that got me onto the *Today Show.*

Arthur the Lyricist

MP: *That's pretty heady.*

Benjamin: Yes, it was. But by this time I think I had established enough friends and other activities that people didn't just think of me as Arthur the Magician. But the thing that I spent most of my time on during my freshman year was writing the lyrics for the big spring musical. Carnegie Mellon had an organization called Scotch and Soda, which had a long tradition of producing an original musical each year. Usually that musical was written or proposed the year before, so that people had a chance to do the writing at least a year in advance. Apparently, that year they didn't have any submissions that were ready to go. So they had the authors present submissions, and one of the authors, Scott McGregor, was still looking for a lyricist and a composer.

I always wanted to write lyrics. If you had asked me in high school, what I wanted to be when I grew up, I might have said a Broadway lyricist. At Carnegie Mellon, I wrote a musical called *Kijé*, based on an old Russian folk tale, that prompted the popular Lieutenant Kijé Suite by Prokofiev. A graduate student had written a "book" for it. A talented freshman named Arthur Darrell Turner wrote the music. It wasn't until a few months later that it was actually selected, but we started writing right away. I was having a lot of fun. Turner could write music, and I could write lyrics. We fought like crazy, but it was a dream come true. The musical was performed on two weekends, with six performances. It was the biggest money-maker that Scotch and Soda ever had. It was a big hit. I had delusions of Broadway.

MP: *How did you find time for classes?*

Benjamin: I did okay. I got As in my math and Bs in my other courses.

MP: *But you were taking the show seriously.*

Benjamin: All my spare time was going into that musical, and I was performing at this nightclub. And I had classes.

MP: *So your life was pretty full?*

Benjamin: I've always been involved in lots of things. Even now, I still have my hand in a lot of different activities.

MP: *Such as poetry?*

Benjamin: I always enjoyed the art of parody, and in the ninth grade I had to write something in Edgar Allen Poe style. The typical thing would be to take a nursery rhyme and write it as if Edgar Allen Poe had done it, but I had decided that since I had always enjoyed "The Raven," I would do something different. So I wrote "The Raisin" (see p. 120).

MP: *Very good.*

Benjamin: That was one of the best things I've written in my life. I think that's how I got the job writing lyrics for the *Kijé*.

The Whole Problem Is Memory

MP: *On the general subject of memory, during your years at Carnegie Mellon, your memory was enhanced.*

Benjamin: Right. I participated in experiments for Professor Chase—I would do calculations, and I would think out loud how it worked. In some experiments he just timed me doing problems. In others he said, "Okay, think out loud as you're doing this" while somebody was transcribing the protocol. Then he gave me a problem that was sufficiently large that I had to slow down.

Some of the processes had gotten so quick by that time, for example, that it was difficult to explain that 56 times 7 is 392. Was it memorized? Was it something I was actually doing? The processes were so fast that they were hard to articulate. But when I was doing very large problems, everything was slowed down, including the simple steps. So that 7 times 50 is 350, plus 42, is 392. So now I know quite clearly what I'm doing. Also, the use of mnemonics, being able to take three-digit numbers and replace them with a single word, expanded my calculating potential because multiplying five-digit numbers and six-digit numbers is not a hard thing to do on paper. It's tedious, but everyone can do it. The whole problem is memory. Most of us can hold only eight plus or minus two digits in our working memory. My working memory was pretty average, and yet, I was able to square four-digit numbers. How was I doing that with only a normal-sized working memory? Chase outlined what I was doing and saw that I never held on to numbers for long. Most numbers get utilized as soon as I computed them.

MP: *In the process of doing this rapid mental arithmetic, properties of numbers were becoming very interesting to you as well.*

Benjamin: Right. I've always been fascinated by numbers. As a kid, I enjoyed casting out nines, and I would check my answers by doing it mod 9 and seeing if it matched up.

The Raisin by Arthur Benjamin

Once upon a day quite cheery,
far from lusterless Lake Erie,
on a beach somewhere southwest of the
    City of Singapore,
'twas a grape that had a notion:
If he'd rest close to the ocean,
he would tan without his lotion,
just by resting near the shore,
and he'd meet up with adventures that
    he'd never dreamt before,
all of this and much, much more.

All of this had happened one day.
It was on a sunny Monday
as he soaked up every sunray going into
    every pore.
Not a gust of wind was breezing,
and the warmth was, oh, so easing
and so beautifully pleasing.
He was filled with joy galore.
"Oh, how I wish," he said, "that I could
    stay here on the shore,
remaining here forever more";

Well, that grape who had that notion,
slowly motioned to the ocean.
He then discovered something that filled
    him with much gore,
for he saw by his reflection
nature made a small correction.
He was further from perfection,
meaning worse off than before.
He had changed into a raisin while he
    rested on the shore,
and that he'd be, forever more.

Well, one day a man in yellow
gazed on at that little fellow
and looked at him and other raisins
    resting on the shore,
for you see his occupation
was to go to this location,
meaning that it was his station,
to pick raisins off the shore.
So he knelt down towards that raisin on
    the beach near Singapore,
picked him up with many more.

All of them were squished together
in a packet made of leather
in a factory southwest of the city,
    Singapore,
and that raisin was a snooper
and he saw a spacious super
massive mammoth monstrous scooper,
scooping raisins by the score,
packaged them in Raisin Bran and sold
    them to the store,
scooped him up with many more.

And that raisin now is well aware
he's in a bowl in Delaware
ready to be eaten by a child not yet four,
and the milk was slowly dropping
and the raisin heard it plopping,
then the raisin heard it stopping
for the milk had ceased to pour.
"Au revoir, sweet life," he cried, "life
    which I truly do adore!"
Quoth the raisin Nevermore.

Benjamin: Give me a four-digit number to square, and I'll outline my method.

MP: *6,743.*

Benjamin: Okay. 6-7-4-3. The first thing I do is to multiply 7,000 and 6,486. Where do these numbers come from? I doubled 6,743 to get 13,486, which separates into 7,000 and 6,486. I do 7,000 times 6,486 in two chunks: 7 times 6,400 is 44,800, plus (7 × 86 =) 602 is 45,402. So that's 45,402,000. At this point I will say "45 million" to get it out of my memory and into the audience's memory.

MP: *I observed that the other night. Go ahead.*

Benjamin: Now I need to store the number 402. Using the phonetic code—4 has the R sound, 0 has the S or Z sound, and 2 has the N sound, so by inserting vowel sounds, 402 becomes the word "raisin." It could be "reason", or "rosin," but "raisin" is perfect. So I say "raisin" to myself once or twice, and now I square 257 (the distance between 6,743 and 7,000) by the same method. I do 300 times 214, which is 64,200, plus the square of 43, which I have memorized (1,849), to get 66,049.

I'll maybe turn 049 into "syrup" (since 0 = S, 4 = R, 9 = P or B) for future reference. I'll say "syrup" a few times then take 66,049, add that to "raisin" (which I translate back to 402,000), and I get 468,049.

What I'm doing algebraically is $A^2 = (A + d)(A - d) + d^2$. I figured this out experimentally in eighth grade.

MP: *In one of the articles that you provided, you talk about, as a young kid, having a Velcro board with the numbers one to ten on it.*

Benjamin: I would think about it . . . multiply those numbers, and I would get ten in so many different ways, I could build up some of the mental muscles. I got to be good at multiplying by one-digit numbers. I was intrigued by my own home address which was 1260 Belrose Road, and 1-2-6-0 is a highly composite number. As a matter of fact, it's half of 2,520 which is the least common multiple of the numbers one through ten. I've always liked playing with numbers.

MP: *But not enough to drive you in the direction of number theory?*

Benjamin: If I had been exposed to it earlier, I'm sure I would have loved it. If I had seen a book like *Power Play,* by Ed Barbeau, I would have eaten that stuff up, and it would have sent me more in that direction. Other than

Figure 7.7 The newly minted Dr. Benjamin with his parents, Larry and Lenore Benjamin, on the occasion of his graduation from The Johns Hopkins University in 1989.

the fact that I liked numbers, I just didn't have much exposure to pure mathematics.

Mathematics and Games at Carnegie Mellon

MP: *But you were likely to get lots of exposure to pure mathematics at Carnegie Mellon.*

Benjamin: Yes, I went to Carnegie Mellon to study mathematics, but another side of me, a very big side of me, was interested in games, backgammon and chess in particular. I was captain of my chess team and played seriously. I studied chess from the seventh grade through twelfth grade. I gave that up when I entered college as well, because I realized that if I were going to go to the next level, that would require a bigger investment of time than I was willing to make. So I put chess off on the side; I could get better at backgammon much more quickly.

MP: *So you got to be very good at backgammon. Do you still participate in backgammon tournaments?*

Benjamin: In fact, I won the American Backgammon Tour in 1997, and, for a while, I had accumulated more points than anyone in the competition's history. So my interest in games has always been there, and that got me interested in probability, operations research, and game theory. I still do a lot of work in that area.

MP: *Let's skip for a minute to your scientific publications. I think some patterns are well established already. Your mode of operation*

with students and teaching is likely pretty well-formed. I am impressed very much that many of these papers were written with students. You might explain that. How does that happen?

Benjamin: Every year, I take one or two senior math majors who want to do a senior thesis with me, and I either give them a problem or a problem area, and we work on them. Often the topic is something that is interesting enough that we send it off for publication.

MP: *That's not easy to accomplish.*

Benjamin: I won't take all the credit. Students at Harvey Mudd College are very good. The problems are my own, and I would work with them, but the students deserve most of the credit. I just try to identify the things that I liked as an undergraduate and what would appeal to me and what still appeals to me. I get a lot of inspiration from the MAA journals. I think they are a good source of problems and ideas.

"I Want to Bring Math to the Masses"

MP: *While at Carnegie Mellon, you must have been wondering about what you were going to do with your life.*

Benjamin: Yes.

MP: *The decision was made at some point.*

Benjamin: The people I looked to for inspiration were guys like Martin Gardner and Carl Sagan. I wanted to be someone who would bring math to the masses, and I could use my entertaining talents to get more people excited about math.

On top of that, I had some teaching experience as an undergraduate. I thought teaching was fun, and it was just like performing except it was better because your repertoire changed every lecture. I even sensed then that if I were a full-time performer, it would get a little bit repetitive after a while.

MP: *Doing the same show again and again.*

Benjamin: I love applause, no question about that, but eventually you start saying, "Yes, I can do this, but what else?" I wanted something more substantial, something more intellectually satisfying. I also felt that if I were going to have any credentials for popularizing mathematics, having a PhD would be essential.

MP: *It doesn't hurt to have the doctorate.*

Benjamin: It seemed like the right path, but if you had spoken to me at the beginning of my senior year of college, I thought I was going to be working at Bell Laboratories in Holmdel, New Jersey, as an operations researcher, and they would pay me to earn a master's degree in operations research. The Bell people interviewed me and wanted me, but that year they had a divestiture, and they weren't hiring undergraduates. So I also interviewed at the National Security Agency (NSA).

MP: *Brent Morris told me about that.*

Benjamin: I met Brent at NSA. Everyone I met while interviewing there said, "You've

Figure 7.8 Benjamin with Brent Morris, another mathematician who practices magic, at the seventy-fifth anniversary celebration of the MAA in Columbus, Ohio, 1990.

got to meet Brent Morris." On my last day of interviewing, I met Brent, and we had a wonderful hour jam session where he showed me his faro shuffle stuff, and I showed him my mental calculation stuff. We have stayed in touch over the years. By the time I heard from NSA, I had graduated from Carnegie Mellon a semester early and had entered the doctoral program at the department of operations research at Cornell. As it turned out, I worked at NSA for a little bit near the end of grad school.

At Cornell I really got a taste of teaching. I was a TA for a stochastic processes class, and that was a job I loved. Unfortunately, because I had graduated a little early, I started in the spring. I was out of sync with the other grad students. I also became painfully aware of the lack of pure mathematics in my background. I had a lot of probability and statistics classes but nothing that required any deep mathematics. So I took some of those courses at Cornell and did well in them, but it was just so much work. I felt as if I was a step behind everyone else who had more pure math in their background. I took my first theoretical analysis class and linear algebra class at Cornell, which I enjoyed, but I decided to start over again. I spent two semesters at Cornell, then a semester back home in Cleveland that spring. While there, I worked as a grader at John Carroll University and came to appreciate what it would be like to teach at a good four-year college. The following fall, I started fresh at Johns Hopkins in the department of mathematical sciences. After that, it was smooth sailing.

MP: *How did you end up choosing Johns Hopkins?*

Benjamin: I was so unsure of what area I wanted, I considered some areas of operations research, discrete mathematics, maybe integer programming, that kind of thing. I was starting to feel that it was discrete math that I really liked, but I didn't think I wanted to delve into a pure mathematics program and have to take graduate courses in topology, complex analysis, and geometry. I wasn't really trained for that. I picked up a lot of that stuff along the way on my own.

Hopkins was one of the few applied math departments that was not oriented along classical applied math lines. It did almost all the applied mathematical areas except

for differential equations and mathematical physics. It was basically operations research, statistics, and combinatorics. It's a smallish department, but it was perfect for me. I got attention from very dedicated caring faculty. They cared about teaching, too, much more so than most universities. Alan Goldman was my advisor from day one. We had a great relationship. Ed Scheinerman was a fresh PhD and had just started on the faculty at Hopkins the year I started as a graduate student. We formed a friendship that has remained strong to this day.

MP: *You're not that far apart in age.*

Benjamin: No, about five years or so. I felt my teaching talent was very much enhanced by watching Scheinerman, one of the best teachers around. The department chair, John Wierman, took good care of me. My first semester there, I did well. They nominated me for an NSF Graduate Fellowship, and I got it. When I interviewed at Johns Hopkins, I was sent around to meet various professors, including Alan Goldman, who is extraordinarily well read and interested in everything. There are tons of library books on his bookshelf. If he ever returns them to the library, they will need to build an extra bookshelf or perhaps a room. When I first met Goldman, he said, "Oh, Art, I understand you're interested in mental mathematics. Are you familiar with this book?" And he pulls out a book called *The Great Mental Calculators* by Steven B. Smith. I said, "Well, yes, I'm Chapter 39." He didn't flinch. He said, "Aha, then you're obviously familiar with it," and he put it away.

MP: *Great story.*

Benjamin: Goldman would try to understand each student and find a problem that fit. He really looked for problems, but in many different areas, and I think I've absorbed a little bit of that mindset myself in how I treat students.

My research interests are not as well defined as those of many others. My publications are all over the place. I find interesting problems, and I just like to work on them. Lately I've had a lot of fun working with Jennifer Quinn, finding combinatorial proofs of Fibonacci number identities.

The problem in my thesis concerned games or maneuvering problems where you want to get objects from one place to another. I proved that if your rules for movement satisfy some simple conditions, then your optimal strategy would be to spend most of your time repeating a few basic patterns of movement. My thesis was called Turnpike Structures for Optimal Maneuvers. It was published in *Operations Research* and was awarded the Nicholson Prize for Best Student Paper from the Operations Research Society of America, which just flabbergasted me.

"A School That Valued Teaching—That's Where I Wanted to Be"

Benjamin: Interestingly enough, when I went on the job market, there was more interest from the big research schools for me than where I really wanted to be—at a school that valued teaching. I got one offer from a big university. The faculty consisted of quantitative modelers, and my thesis was very interesting to them. While I was interviewing at the school I met a young professor, and I noticed on his bookshelf a trophy that the university gave for outstanding teaching. I said, "Hey, that's really great." And he said, "Yeah, that and 50 cents will get you a cup of coffee around here." I said "Would you trade that in for a publication?" He said, "In a heartbeat."

He said in terms of what it was worth as a faculty member there, it would be substantially more beneficial to have one more publication than that teaching award. I just didn't feel that I could be happy in a place where one of my strong suits was not valued. The dean told me, "To make it here, you're going to have to spend a couple of days a week with your door closed where you're in your office reading and writing and grant-getting." To me that sounded like work.

MP: *A job.*

Benjamin: It sounded like a job. That's not what I got a PhD for. Whereas Harvey Mudd sounded like a lot of fun. To be sure, I put a ton of hours into Harvey Mudd. But they are fun hours. And my door is open, and students come by. I can be talking to students several hours a day. My chairman and dean can walk by my office, and they will think that I'm doing a great job. They can look at all the interaction I'm having. At many other places, that would not be considered a very smart use of my time.

MP: *So you found a place where you wanted to be.*

Benjamin: Absolutely. I'm very happy there. The department is rebuilding. HMC is a relatively new college. The school was established in the late 1950s, and almost all the faculty were hired in the 1960s, so we have seen a recent wave of retirements. At one point there was a twenty-five-year age gap between me and the next oldest faculty member. When I went up for tenure, I was the first person to do so in twenty years in my department.

MP: *Wow!*

Benjamin: So now it's a brand new department.

MP: *A young department.*

Benjamin: Three years ago, the median age in my department was 63. Now it's 36.

I am at the median, and I may be above it. And in a few years, I may be the only person here who was hired when I started in 1989. It's exciting, and with my amazing department chair, Michael Moody, we're hiring great new faculty. Our students are very satisfied. About 40 percent of our students go on to get PhDs. In the last ten years, we've managed to get a third-place team finish on the Putnam. Last year we were ninth. Last year, we had five student publications in refereed journals.

Marriage, Children, and the Future

MP: *That's great. Now there's another dimension or two of your life here—your wife, Deena.*

Benjamin: We met at Johns Hopkins when I was a graduate student. She was an undergraduate. We got to be very close friends our last year at school. Then we went our separate ways. Immediately thereafter, I went off to Harvey Mudd, and she went off to the University of North Carolina in their operations research program. She was there for two years and got her master's. When she was looking for a job, I suggested that she consider California. A year later, we were engaged, and the following year we were married. She started working at the RAND Corporation and was there for five years. She now does computer contract work, technical writing, and soon motherhood.

Figure 7.9 Art and Deena Benjamin relaxing at home.

MP: *What do you hope for your child? The gene pool is sort of stacked for this kid.*

Benjamin: Well, our plan is to do what my parents did, which is to expose the child to everything and let him or her find his or her way and find things that are enjoyable. The thing that bonds my brother, my sister, and me together, as I said before, is our love for theater and performance. That has helped all of us in our lives, but no person should go on the stage if he doesn't want to go. I'll certainly present mathematics in the best light, but I'm not going to force it down their throats. I'm not going to try to turn our children into professional mathematicians.

MP: *For many faculty, the classroom is their stage. It sounds to me as if it is an important stage for you, too.*

Benjamin: Oh, yes. I think every lecture should have a little bit of relevance or elegance because that's what students respond to.

MP: *But you have every intention of continuing your magic and your mentalist activities?*

Benjamin: It's a part of me. Although I fully expect that with the arrival of a child my life will be altered significantly.

MP: *At least for a few years.*

Benjamin: Yes, I'm going to spend a lot of time with the baby. Another long-term goal is seeing Harvey Mudd College and the mathematics department there blossoming and reaching its full potential. I think the Harvey Mudd department has a good

balanced view of teaching and research, and the students really benefit from that.

MP: *Would you ever want to be chair?*

Benjamin: No, no. I do not want to be a chair, partly because I know myself well enough to feel that that would not be my strongest suit. There are aspects of being a chair that play into my weaknesses. I've had experiences in other organizations, and I know what I'm good at. I'm a good person to work with a chair and be put in charge of certain types of initiatives. And I'm very content with what I'm doing.

MP: *What other goals do you have?*

Benjamin: I have a goal of bringing mathematics to the masses. A few years ago I did develop educational material on doing mental math. I created a book and videotape and audiotapes, which were mass-marketed for a time with an infomercial. That was sort of a risky thing to do, especially around the time I was going up for tenure. That has worked out fine. It was based on the book, and it's been great. Lots of people have learned how to do math in their head through these programs. That was partly a dream come true.

I haven't reached Carl Sagan or Martin Gardner proportions yet, but it did bring mathematics to a lot of people!

Postscript: Professor Benjamin, a decade after the original interview, has added to his list of accomplishments the co-editorship of the student journal, *Math Horizons*, has two daughters, Laurel and Ariel, and, yes, has even had a term as department chair at Harvey Mudd College. He has created two DVD courses for the Teaching Company on the Joy of Mathematics and Discrete Mathematics. In 2006 the American Mathematical Society gave its first award for an Outstanding Mathematics Department to Harvey Mudd College.

Eight

Dame Mary L. Cartwright

James Tattersall and Shawnee McMurran

Mary Cartwright was born December 17, 1900. She matriculated at St. Hugh's, Oxford and was awarded a DPhil from Oxford in 1930. Her thesis advisors were G. H. Hardy and E. C. Titchmarsh. During her career she made important contributions to the theory of functions and differential equations. She is particularly well known for her work with J. E. Littlewood on van der Pol's equation. Dame Mary was elected a Fellow of the Royal Society of London in 1947. She received the Sylvester Medal of the Royal Society and the De Morgan Medal from the London Mathematical Society. Cartwright served as president of the London Mathematical Society and of the Mathematical Association (British). She received honorary degrees from the Universities of Edinburgh, Leeds, Hull, Wales, Oxford, and Cambridge and became a Dame of the British Empire in 1969. She was Director of Studies in Mathematics and Mistress of Girton College and Reader in the Theory of Functions at Cambridge University. Dame Mary died in Cambridge, England, on April 3, 1998. The following interview was conducted in person and through correspondence during the period 1990–1994.

MP: *Would you tell us about your family?*

Cartwright: I am descended from the Cartwrights of Aynho in Northamptonshire. Richard Cartwright purchased the manor house and its grounds in 1616. His son John, who married the daughter of the attorney general to Charles I, remained a Roundhead and stayed in London during the Civil War. During the War the Aynho Park manor house was occupied by Cavalier soldiers, and John's mother was imprisoned in Banbury. John's only son William married Ursula, daughter of Fernando Fairfax. After their major defeat by Thomas Fairfax at the Battle of Nasby, the returning Cavalier soldiers burnt most of Aynho Park.

My great-great-grandfather married Mary Catharine Desaguliers, whose grandfather, J. T. Desaguliers, came over from France with his father, a Huguenot pastor, after Louis XIV revoked the Edict of Nantes in 1685. J. T. Desaguliers became Curator of the Royal Society and wrote books on physics. His son became a general in the army and brought back stones with inscriptions

Figure 8.1 Dame Mary Cartwright, 1965.

praising Louis XIV, which were inserted in the wall at Aynho Park. My great-grandfather, William Randolph Cartwright, lived in Rome for a time and then Florence, where he was a correspondent for the *Spectator* magazine. He returned to England and served as a liberal MP for Oxfordshire for more than forty years. He was married twice and had thirteen children, eight sons by his first wife and five sons by his second wife, Julia Frances née Aubrey.

My father's father was a colonel in the Grenadier Guards and served later as an MP for South Northamptonshire. My parents married in 1894 and lived at Church Cottage in Aynho, where my father served as curate

to his uncle the rector. My father read history at Christ Church, Oxford. My mother was Lucy Harriet Maud Bury (rhymes with berry), and her father, Edward Bury, was a barrister who died in midcareer.

Early Life

MP: *Would you give us a feeling for what your life was like in the early 1900s?*

Cartwright: When I lived in Aynho, from 1900 until my father's death in 1926, only four houses had tap water and proper drainage: the main house at Aynho Park, the grammar school, the estate agent's house, and Church Cottage, where we lived before my father became the rector. Even these had rather inadequate plumbing. We used rainwater from the roof for sanitation, and sometimes water had to be pumped from a well. We left Church Cottage before I was six, but I remember it had a downstairs water closet.

MP: *Did you have any brothers or sisters?*

Cartwright: There were four other children in my family: John, Nigel (Walter Henry), Fred (William Frederick), and Jane. John Digby was born in 1895, and Nigel in 1897. John, from what my mother told me, had a great deal of charm from the time he was quite small onward and was a natural leader but very poor at learning to read and, later, at all schoolwork. He had to have special coaching to get into Sandhurst and became a second lieutenant in the 2nd Battalion of the Durham Light Infantry, the regiment of my father's eldest brother Henry Aubrey. Nigel was larger and, as a child, more awkward but very much cleverer than John or I. John and Nigel got on very well together, and I, of course, admired them and always regretted having to turn back on Sunday walks because I could not go as far. Nigel was interested in carpentry and building with toys and bricks and good with boats. I watched his building and tried a little. I did not get very far; no ideas of my own was the trouble.

Nigel matriculated at Christ Church, Oxford, with a view to reading classics after the War [World War I]. He became an officer in the 2nd Battalion D. L. I. and was killed in 1917. John was also killed in the War. Neither has a known grave, but their names are inscribed on the Menin Gate at Ypres.

MP: *Tell us about your younger brother and sister.*

Cartwright: I was born in 1900, my sister Jane in September 1904, and my brother Fred in November 1906. Jane was a good deal larger than I, taller and broader, and grew a lot when she was quite young. This was embarrassing for her because, like John, she was slow to read and found spelling and all school work difficult. I wonder now whether she was dyslexic. Fred was much quicker but not outstanding until he got a job with a steel firm and got transferred from the London office to work in South Wales. As a child he used to do interesting things with a Meccano set and in carpentry, but at his preparatory school his reports varied up and down. When I coached him in Latin, he was moved up a form, which upset his mathematics for a bit. However, he settled down enough not to have serious difficulties.

Conic Sections Intriguing

MP: *Tell us what you remember about your early education.*

Cartwright: I tagged along after John and Nigel whenever I could. John went to school first, and one of the few things I remember before I was six is playing with Nigel in a yew tree near the village infant school. The mistress, daughter of the headmaster, gave

Nigel a box of colored chalks and asked us to go away, which we did. John and Nigel were taught by the eldest daughter of an estate agent, and the schoolroom was built of corrugated iron leaning against the house. I think I must have been allowed to attend their lessons and given some help. I remember having a very old-fashioned chunky book with short words and black line illustrations. Much later I found a reading book of mine with words of five letters dated December 1905.

In the days at the rectory and early years at Leamington High School I avidly read *Little Folks*, a magazine for children published by Bella and Leonard Woolf. My mother took a lot of trouble to get educationally improving toys for us. There was an organization, the Parents National Educational Union, that helped provide materials and books for education at home. They offered a correspondence course to help governesses. My mother knew someone who belonged and got ideas and material from them. In particular, bricks, carpentry tools, and Meccano for my brothers, and for me a set of wooden models: sphere, cylinder, cone, and perhaps others. It was sections of the cone, the ellipse, hyperbola, and parabola, that intrigued me the most. Unfortunately, the little metal peg at right angles to the ellipse did not fit tightly enough to hold the top, which always pointed downwards as soon as I let go. The angle of the peg should have been slanted upwards. There was a geography book, but it was deadly dull, and I could not learn exports and imports without any explanations or pictures. There were also little arithmetic cards from the village

Figure 8.2 Portrait of Dame Mary that "misses the warm sense of humor and sympathy that her friends and students knew."

school. I don't know what ages they were meant for, but I suspect that I was doing thirteen- to fourteen-year-age cards before I was eleven.

MP: *How rigid were social distinctions in England in the period just before the First World War?*

Cartwright: The country gentry did not mix socially with the farmers nor with people who actually lived in Banbury, the nearest town of any size. Brackley and Deddington were small enough for people in some of the larger houses to be acceptable. The children of the gentry could not possibly be sent to the village school, so the boys were sent to preparatory school from age eight or nine to age thirteen or fourteen and then to public schools until eighteen or nineteen.

At this time, 1907–1910, there were various schemes for improving the education of girls. A few girls' boarding schools comparable with boys' schools like Harrow, Winchester, Rugby, and so forth, had been started, but the parents who lived in the great country houses usually preferred to have a resident governess, which was expensive. A friend about eight years older than I said once, "Oh yes, we had governesses. I got through nine."

MP: The first secondary school you attended was Oxford High School. What do you remember about that experience?

Cartwright: Oxford was only eighteen miles away from Aynho by train, and the trip was familiar to me from quite early days. I had attended dancing classes in Oxford at the Randolph Hotel with about fifteen or twenty other children. We learned the polka, the waltz, and sang nursery songs with movements to the music. Mother bought a very small pink paperback in the series *Books for the Bairns* for me to read on the train. Nigel and I were often sent to stay with Aunt Rose, mother's next eldest sister. Aunt Rose's husband, Warner, was a tutor at Christ Church. I am afraid that I did not like staying with Aunt Rose. There was far too much white paint, and three mats to wipe your feet on at the front door, one outside and two inside. We had family prayers with the servants, except of course the cook, as at home. The gas fire in the breakfast room fascinated me, as did the stuffed duck-billed platypus on the piano. I think that the Oxford High School interlude was to get me used to being

with children of my own age because it was so difficult to make contact with any in the neighborhood at Aynho.

Relativity and School Geometry

MP: You then went to the Leamington School. What kind of mathematical foundation did it provide?

Cartwright: I returned to school the following autumn to a higher form with my cousin Christine. We began Latin and the simplified Euclidean geometry which was done in all schools then, using a book by Hall and Stevens. It was soon superseded by the Godfrey and Siddons. Shortly thereafter, the Mathematical Association produced a report saying that proofs by superposition that triangles are congruent under certain conditions were invalid since the theory of relativity implied that things moving from one place to another at great speeds do not necessarily remain the same size. The report placed me in a very awkward situation several years later at the Wycombe Abbey School when I found myself teaching Siddon's daughter using her father's book and was obliged by the head of the department to tell her to cross out large portions of the text.

I liked geometry at Leamington, but there were times when I thought we would never get beyond multiplication of decimals. Mrs. Pochin, our teacher, had attended Newnham College, Cambridge. On one occasion she held forth about prime numbers and how difficult the problems about them were. I felt they were not for me. I wanted to solve problems and not get stuck. As it turned out, I never produced any new result in the theory of numbers but only toyed with the field in order to see what the difficulties were like.

MP: *In the fall of 1916 your parents sent you to the Godolphin School in Salisbury, Wiltshire, about twenty-five miles north of Bournemouth. Isn't that where you were first exposed to the calculus?*

Cartwright: The Godolphin School was for girls only, many of whom were daughters of doctors, clergy, missionaries, and farmers. The residence accommodations were almost unbearably primitive and old fashioned. There were day girls as well as boarders, and some daughters of the owners of big shops in the town. It was during the war, so we never went into City Center except to the Cathedral Services on Sunday afternoons at 3:30 p.m. Each group walked there in crocodile fashion, that is, two and two with the house mistress at the back. It was rumored that Australian soldiers on leave might be dangerous, and so we went for walks in fours when it was not possible to play games. I had lost time by changing schools but was put into a small advanced section for mathematics taught by Miss E. M. Hancock. She had learned mathematics by reading tutorial books for the external University of London BA degree. I first learned about the calculus, analytical geometry, and uniform convergence from her. It was Miss Hancock who encouraged me into reading mathematics at Oxford.

College

MP: *Oxford had several colleges open to women. Why did you choose to matriculate at St. Hugh's rather than Lady Margaret Hall, Somerville, or St. Hilda's?*

Cartwright: My cousin, Cecily Ady, was a Fellow at St. Hugh's and taught history. Her mother, Julia, née Cartwright, wrote well-known semipopular books on the history of Italian painters, and by her writing, which included a multitude of cheaper books and pamphlets, earned enough to pay for Cecily's education. Julia, who resented not being allowed to learn Latin with her brother Chauncy, was taught German and other subjects considered suitable for girls. At the time, in order to enter Oxford, one had to pass an external examination in several subjects including arithmetic, the classics, and modern languages. My last year at Godolphin, I qualified by doing the Senior Cambridge Local and completing course work in Latin prose and Greek, which were then essential. I took the entrance examination to St. Hugh's in the spring of 1917 in mathematics, which consisted, for me, of papers in pure maths only. I could not attempt the scholarship examination, which included statics and dynamics, because I had done no applied maths. Miss Hancock taught me a little science in the Summer Term and even rigged up an experiment to measure gravity. I cannot remember the precise details, but the experience put me off all experimental work. I think perhaps with respect to the sciences she did not prepare my mind enough. She was very good, but not successful with all the girls. Most upper-class girls were brought out, presented at Court, and spent a year or two attending all the fashionable functions, dances, racing, and so forth. They did not usually go to university.

MP: *What did you do during the long vacations?*

Cartwright: After my mother died I stayed with my cousin Di for a week and then again after the summer term every year until she gave up the house which she had built. It was thatched with Norfolk reeds and was so flimsy that it was a fire hazard, but she loved its rural appearance festooned with roses, wisteria, and jasmine. She would not let her own gardener cut anything because he would have been too drastic. But I cut a little the first day and a little more the second day. Every year she took me to a Shakespeare play at Stratford-on-Avon about thirty miles away. I remember *A Midsummer Night's Dream*, *Much Ado About Nothing*, *The Tempest*, *Macbeth*, and *Henry V*. That is to say I remember something which makes me sure that I saw them. For example I thought that Prospero's crown with lighted candles was dangerous and that the business with Banquo's ghost was not well managed. We always had seats in about the third row where you could see the actors' saliva when they were especially impassioned. Such are the memories which remain! The thing that pleased me most was that Di always booked lunch for us at a small hotel very near the theater.

MP: *How difficult was it for you as a woman to read mathematics at Oxford?*

Cartwright: I came up knowing that I was ill prepared. The honors maths course at Oxford consisted of passing the Mathematical Moderations, which can be taken in two years, or only one if you are well prepared, and then another two years of work for Finals. I had to take two years over Mods [preliminary examinations taken at Oxford two years before Finals], which made any university scholarship out of reach for me. At that time there was no money at all at St. Hugh's for postgraduate work. In October 1919 the University was flooded with men released from the army and not nearly enough people to teach them. I managed to get copies of notes from J. W. Russell, who taught women from other colleges, but he would not undertake me and my two fellow students from St. Hugh's. He did an enormous amount of teaching and lecturing for Balliol and wrote two books on geometry. The year before I took Mods, many men returned from war service and took the examination. An unusually large number of firsts were given, but in my year there were only three or four. I got a second and I had to choose whether to continue to Finals or to change to some other subject such as history. I decided to continue in maths largely because history seemed to entail longer hours of work. What really hooked me on mathematics was the calculus of residues. I should have mentioned that the Oxford system then, and for a long time thereafter, required students to go to a particular set of lectures given in the Colleges and have one hour per week of supervision alone or in pairs or possibly more, for which written work was required. I did not get very much out of those hours alone except at first. My first supervisor, L. J. Rogers, a good mathematician as I later found out, had retired from a professorship at Leeds but seemed bored. I was sent to J. W. Nicholson for one term of supervision before Finals and found him good. Unfortunately, it was in dynamics, which was not a subject that appealed to me at the time.

"I Owe My Career to Morton"

MP: *How did you first meet G. H. Hardy?*

Cartwright: Near the end of my third year at Oxford, one of the most important events in my mathematical career occurred at a party on a barge in Eights Week [the main annual rowing event held at Oxford in May on the Isis]. The men's colleges had barges on the Thames that their rowing men used for changing. When the races took place there were parties on the roofs of the barges. Mildred Cousens had asked me to go with her to a party. The chaperone rules were very restrictive, and we avoided having to ask permission if possible. So I think it was probably okay for such a party if there were two women and two men. One of Mildred's friends had asked V. C. Morton, later professor of mathematics at Aberystwyth, to the party. At one point during the evening Morton advised me, if I was serious about mathematics, to read E. T. Whitaker and

G. N. Watson's *Modern Analysis* and to attend Professor Hardy's class, which took place once a week over the Holywell gate of New College Friday evenings from 8:45 p.m. to 11:00 p.m. I owe my career to Morton.

I read *Modern Analysis* during the following long vacation and got special permission to attend Hardy's class. Hardy's class consisted of a talk of about an hour always beginning fifteen minutes late. Afterwards we had tea and biscuits and talked about mathematics and mathematicians. Apparently, I impressed him very favorably for he was a Finals examiner when I got a first. One paper was so hard that no one seemed able to do more than twenty minutes work on the questions. Hardy took immense trouble with his students whether they were good, bad, or indifferent. Once, when I had produced an obviously fallacious result, Hardy said, "Let's see, there's always hope when you get a sharp contradiction."

Hardy's manners were peculiar. If he walked along a street he looked at the ground, only looking up occasionally. I think that if he saw someone he knew he would deliberately look the other way.

MP: *What did you do after obtaining your degree from Oxford?*

Cartwright: I taught mathematics for one year at the Alice Ottley School in Worcester. When they had to economize, I left, and I went to the Wycombe Abbey School, an expensive boarding school in High Wycombe, where I had to teach mathematics exactly in the way the senior mathematics mistress said. There was virtually no chance to experiment at Wycombe Abbey. Eventually I felt I had to try something else. I approached Professor Hardy in January of 1928 and arranged to attend his class for research students and work under him.

Being a Student of Hardy

MP: *Would you describe what it was like to be a student of G. H. Hardy?*

Cartwright: From 1920 to 1931 Hardy was the outstanding influence in pure mathematics at Oxford. I felt that I had joined a rather very special group when I began to attend Hardy's class at New College in January 1928. He had just ceased to be secretary and was beginning his term as president of the London Mathematical Society. He was obviously strained and working very hard. Looking back, I think that we knew that we were studying under a very great man who had not yet been recognized fully. The members of the class included Gertrude Stanley, who had finished her course and was a lecturer at Manchester, and John Evelyn, a direct descendent of the famous diarist. Evelyn came only occasionally, and I assumed that he had obligations associated with his ancestral home in Wotton-under-Edge. Also there was E. H. Linfoot, L. S. Bosenquet, and an Indian named Vijayarhagavan who was so highly regarded by Hardy that he was selected to lecture when Edmund Landau paid a social visit. A Canadian, Frederick Brand, and I were also selected to speak

Figure 8.3 G. H. Hardy

before Landau. Gertrude Stanley said to me early on, "We all go to Hardy's Monday class for undergraduates as well, but we don't speak." In his seminar one evening in 1928 Hardy gave us a list of problems. He was amazed when I did one completely by straightforward contour integration. My work appeared as an appendix to his book on divergent series.

MP: *Hardy was in the United States while you were working on your thesis. Did that cause you any problems?*

Cartwright: From October 1928 to May 1929 Hardy exchanged jobs with Veblen of Princeton, a geometer. E. C. Titchmarsh was put in charge of me, and his suggestions suited me well. Titchmarsh had a job at University College, London, and also a Research Fellowship at Magdalen College, Oxford. We used to meet in his rooms in Trinity. He was a wealthy man with an estate in Northumberland. Tea was served on a silver tray from a silver teapot. Hardy returned from the United States in May 1929. I thought he looked very much better, and quite relaxed, but his hair had turned white. After his return I completed my DPhil. I successfully applied for a Yarrow Research Fellowship at Girton College, Cambridge. It was said that the degrees of PhD and DPhil at Oxford were only created about 1920, "for Americans." In the early 1920s men who were interested in a research position job at Oxford were told to be content with the traditional BSc and not to go for a DPhil. In Cambridge and elsewhere jobs depended entirely on research and personal recommendations. The numbers involved were so small that everyone had personal contacts if they were any good at all, and this was true until after the Second World War. Very few people were graduate students at universities outside Oxford, Cambridge, and perhaps two or three of the London Colleges. Edinburgh developed a graduate school of modest kind among its junior lecturers, most of whom had no previous research experience. At Cambridge, a graduate scholarship would enable a good mathematician to write a dissertation which he could submit for a prize fellowship at his college. For example, at Trinity, to the best of my knowledge, there were no restrictions or duties, and the fellowship could be held for four years, after which the man's prospects for a college lectureship were good if his research work was good. I regret to say that my impression when I began research

was that, in general, less qualified men were tolerated and employed quite a lot in the University of Cambridge, which eliminated some quite good women from Cambridge.

Hardy Heckles Littlewood

MP: *How did Hardy's seminars at Cambridge differ from those you attended at Oxford?*

Cartwright: When he returned to Cambridge, I asked Hardy if he would be offering a seminar similar to the evening sessions I had enjoyed at Oxford. He replied that he would probably come to some arrangement with Littlewood. Soon after, the Lecture List announced a Hardy-Littlewood class. While Littlewood was speaking at the first class, Hardy came in late, helped himself liberally to tea and began to ask questions. It seemed as if he were trying to pin Littlewood on details, whereas Littlewood was trying to illustrate the main point while taking the details for granted. An irritated Littlewood told Hardy that he was not prepared to be heckled. I don't recall them ever being present together at any subsequent class. Thenceforth, Hardy and Littlewood alternated classes. Littlewood often did all the speaking on his turns. Hardy often invited others to speak during his classes. Eventually, Littlewood ceased to participate, even though the class continued to be held in his rooms. Consequently the class became known as "the Hardy-Littlewood Seminar at which Littlewood was never present." Thereafter, Littlewood intermittently held

Figure 8.4 At Cambridge in 1938 at the Hardy-Littlewood Seminar, the young Ralph Boas, armed with an unobtrusive and quite secondhand camera, managed to catch a number of mathematicians unaware before he was "apprehended" by Mary Cartwright. In this photo, Mary Cartwright is seated next to G. H. Hardy.

Figure 8.5 Over tea, after a lecture at the Hardy-Littlewood Seminar, Mary Cartwright continues in conversation with another mathematician.

his own unlisted lecture class, which Hardy recommended to a number of his students. R.E.A.C. Paley was by far the brightest of the group that worked with Hardy and Littlewood.

MP: *Do you remember any of the other women mathematicians from your early days at Cambridge?*

Cartwright: I remember Frances E. Cave-Browne-Cave, who was bracketed with the Fifth Wrangler on the 1898 Cambridge Mathematical Tripos. Hardy was Fourth Wrangler that year. Before becoming a Lecturer in mathematics and later Director of Studies at Girton, she [Cave-Browne-Cave] did statistical research with Karl Pearson at University College London. She was a Fellow of the Royal Astronomical Society and published papers in barometric statistics. Her sister, Beatrice, also read maths at Girton but took a second class on the Tripos. Frances Cave[-Browne-Cave] told me that once, while they were being taught by William Young, he kept tilting his chair until it slipped and he went under the table. With great difficulty she and the others refrained from laughing. He then got up and all he said was, "Take out a fresh sheet of paper." Young married Grace Chisholm, whom he had tutored at Girton. They went to live abroad, do research, and raise a family. They had a very successful mathematical career.

MP: *You have more than ninety mathematical publications chiefly on the theory of nonlinear ordinary differential equations, functions of a complex variable, and topology. Which of your articles are of significant importance to you?*

Cartwright: My work on directions of Borel spreads over four issues of *Comptes Rendus* was mentioned by Georges Valiron in his

Figure 8.6 Dame Mary lecturing.

lectures at the Zürich International Congress in 1932. I completed the work in two or three papers in the *Proceedings of the London Mathematical Society* in the 1930s leaving one theorem announced in the *Comptes Rendus* unproved. During the 1930s, E. F. Collingwood refereed nearly all my papers on entire integral functions in the *Proceedings of the LMS* and helped me a lot. Collingwood gave a proof of a theorem proved by Ahlfors while Littlewood was lecturing on multivalent functions. Having seen an early version of the paper on inequalities in the theory of functions, which appeared in the

Mathematische Annalen in 1935, I applied Ahlfors's result to multivalent functions.

Remaining at Cambridge

MP: *What factors played a role in your decision to remain in Cambridge?*

Cartwright: Hardy and Littlewood recommended that my Yarrow Fellowship be extended and I be appointed an Assistant Lecturer at Girton. In October 1949 I became Mistress of Girton. While at Cambridge I served on many committees and was always a halftime university lecturer until I was made a Reader. It was probably being head of a college and not having enough time for research students that stopped my promotion to professor.

MP: *Why did you accept the top position at Girton knowing that it was going to interfere with your research?*

Cartwright: I myself found it virtually impossible to refuse to be elected Mistress of Girton. On the other hand, because of that position and others I have held, I felt favored and preferred to men on many occasions. In particular, on a number of occasions I have been asked to represent the London Mathematical Society, the Cambridge Philosophical Society, and Cambridge University. I served as chairman of the Faculty Board at the university, two terms of office on the Council of the Senate, the senior body at Cambridge University, and on the Council of the Royal Society. These were not salaried jobs. I paid my own way to attend numerous conferences, whereas

a married man, or woman, probably couldn't afford it. Some of the most outstanding male mathematicians did a prodigious amount of work in their heyday. I could not compete.

MP: *Tell us about your research students at Cambridge.*

Cartwright: My research students were not very numerous. Being Mistress of Girton was too demanding and time consuming. I refused to take on extra work if I could help it. Littlewood took over Chike Obi, a self-taught Nigerian, and pushed him through a PhD. Obi is now retired from Lagos University and is a great man in his own country. Sheila Scott MacIntyre married before taking her PhD. Hilary Shuard did not get a PhD, but is prominent now in the teacher training world. Barbara Maitland became a

Figure 8.7 J. E. Littlewood.

Lecturer at Liverpool. Her career was upset by the Second World War, but she managed to survive. James Ejeito served as vice-chancellor of Nsukka University of Nigeria. Carl Lindon taught at University College of Swansea. I examined Noel Lloyd for his PhD and have maintained contact with him. W. K. Hayman, FRS, my star pupil, won the LMS Berwick Prize and taught at Imperial College before moving to York University.

Working with Littlewood

MP: *You collaborated with J. E. Littlewood on several papers. Describe what it was like to work with Littlewood.*

Cartwright: I first met Littlewood in June of 1930 when he came to Oxford to examine me for my Doctor of Philosophy degree. Later at Cambridge I attended a series of courses he offered on the theory of functions. Our work together began with a request in January 1939 from the Department of Scientific and Industrial Research for help in solving certain very objectionable looking differential equations occurring in connection with radar. At the time I had been in the habit of showing Littlewood anything which I thought would interest him. I read a summary by Balthasar van der Pol of work up to 1932 and studied the references back to work by Edward Appleton and van der Pol in 1920. I told Littlewood of some of the nonlinear problems which seemed to arise. We translated problems which were suggested by radio waves and oscillations into dynamical problems. He solved or half-solved several of the problems

and suggested methods to handle the other problems. There was an odd phenomenon occurring with an odd nonlinear function mentioned in a short letter to *Nature* where van der Pol suggested that it occurred with a similar even nonlinear function. That was the only case that really interested Littlewood. He gave me the impression that this was an exciting problem and he would be interested in any significant progress I could make towards its solution. Meanwhile, in one of Edward Collingwood's classes, I had learned of Ahlfors's distortion theorem. While taking a bath one night, I saw how to apply it and Montel's normal families to the problem. When I later settled down to examine the problem more carefully, I fell into the "usual trap." The proof for the case $p = 1$ was known, and I tried to prove the case for $p > 1$ using a minor modification of that proof rather than my original idea. I sent this attempt to Littlewood. In reply I received a note with a sketch of a snake signifying that there was an obvious error in my proof. Returning to my original idea, I sent a new proof to Littlewood using Ahlfors's work. Some time later, while punting with the Vice-Mistress of Girton on the River Cam near Trinity College, I spotted Littlewood on a nearby bank. As I had

Figure 8.8 Littlewood's snake.

received no reply regarding my revised proof, I asked him if he had read my manuscript. He said, "Have I got to read all that?" I convinced him that he should, and when he had he was favorably impressed with my work. We collaborated in proving all solutions bounded, without which all other work was invalid. After a few years I got fed up with incessant changes, and we separated. I left it to him to complete the proofs of the most difficult problems, and I picked up the other bits and pieces.

MP: *How did your collaboration with Littlewood compare to that of Hardy and Littlewood?*

Cartwright: My collaboration with Littlewood was unlike his with Hardy in one respect. He would not let me put his name to any paper not actually written by him. I had to say it was based on joint work with him. There was one exception, a fixed-point theorem on which he gave way on condition that it was checked by someone acceptable to him. The paper was published and later referred to in a subsequent article by M.H.A. Newman in his presidential address to the London Mathematical Society.

Littlewood was always more ready to talk out of doors. He never discussed problems with me at a blackboard, but he would draw imaginary figures on a wall as we walked and talked. I ran into him quite unexpectedly once on the Madingley Road. As we spoke one lens fell out of his glasses. It was in the days of clothes rationing and the knees of his trousers were torn and mended, one by a safety pin. He apologized, adding that he had not planned to call on a lady that afternoon.

Once we began to collaborate, nearly all of our collaboration was done by letters with occasional short discussions of particular points. I was occasionally able to catch him on the telephone after late tea and before his early dinner. I should mention that Littlewood suffered, so he said, from depression from 1931 until about 1956. All my work with him was done during his worst years. Littlewood was a very great mathematician. He once told me, and I agree, that in true collaboration the authors do not know who had which idea first. I believe that his work with Hardy is more important than the work done by either of them separately. I also think that he could not have achieved as much if he hadn't let Hardy write up the final versions of much of their joint work.

Solomon Lefschetz

MP: *In 1949 you were a consultant to the United States Navy. How did that come about?*

Cartwright: Nicolai Minorsky's book had come to my notice. It was a translation of a Russian work judged to be relevant to war-related problems and named [Solomon] Lefschetz as head of the project under the Office of Naval Research. I wrote to Lefschetz at Princeton telling him about the nonlinear differential equations that Littlewood and I were working on. From January to June 1949 I was in the United States. I spent three weeks in January at Stanford with John Curtiss and one week in Los Angeles, where Minorsky worked. From February to June I had an office in Fine Hall at Princeton. My

Figure 8.9 John Tukey.

engineer but lost both hands in an accident and turned to maths. I believe that he did outstanding work in topology during his work in some Middle-West job. In my opinion he was no good at differential equations, but of course he had a feeling for the applications and general background and liked organizing things. While at Princeton, I gave a seminar, and my lectures were published in the Princeton Series of the *Annals of Mathematics*.

MP: *Would you comment on your membership in the Royal Society of London?*

Cartwright: Women have been admitted to membership in the Royal Society since 1945. Almost all such women work in the biological sciences. I was elected in 1947. The second female mathematician, Dusa McDuff, was elected in 1996. Very few women have been elected to the Royal Society who combine teaching and research; still fewer combine marriage and the Royal Society.

agenda was arranged by Mina Rees. Officially, I was a consultant under the Office of Naval Research at Princeton because the university didn't have women professors there, except one without tenure who taught Russian. I got on well with all those I met. Bochner was not on speaking terms with Lefschetz, so I never saw him. John Tukey told me that I ought to put my feet on the table in the seminar room, but, wearing a skirt and not trousers, I refrained. I listened in on a PhD exam of a good candidate who passed, but Lefschetz said that one of the candidate's statements was wrong. The candidate came to me with a reference indicating that he was right.

I learned that if Lefschetz stopped asking questions for five minutes he was asleep. On at least one occasion, he asked me to explain what a visiting lecturer had said. As you probably know he was trained as an

Figure 8.10 Dame Mary Cartwright with Lord Butler and Lord Adrian, escorting Lord Adrian to his installation as Chancellor of the University, June 1968.

MP: *Would you leave us with a bit of your philosophy?*

Cartwright: Before Littlewood and I started collaborating, except for his work on anti-aircraft fire during the First World War, there seemed to be a deep divide between pure and applied maths. I am of the opinion that the dividing line between strictly abstract reasoning and thinking in terms of real-world problems is by no means clearly defined. A number of major developments in pure maths were first thought out in terms of real-world situations. Nevertheless, various aspects of mathematics attract different people who are left cold by others. I have obtained immense satisfaction from doing mathematics. My method was to polish and polish and, when the work had been published, to destroy all earlier versions. Mathematics is a young person's game mainly because major advances in the subject come from approaching problems from a slightly different perspective than previously adopted. These types of ideas often come in the course of learning a subject for the first time. Younger mathematicians have slightly different backgrounds, find it easier to learn new subjects, have been trained in newer methods in allied fields, and are not afraid to experiment with new techniques. It is easier for older mathematicians to let ideas and methods become fixed and to fill their time with other interests.

REFERENCES

Cartwright, M. L. "Non-linear vibrations: A chapter in mathematical history." *Mathematical Gazette,* 36 (1953): 81–88.

———. "Mathematics and thinking mathematically." *American Mathematical Monthly,* 77 (1970): 20–28.

———. "Some points in the history of the theory of nonlinear oscillations." *Bulletin of the Institute for Mathematics and Its Applications,* 10 (1974): 329–33.

———. "John Edensor Littlewood, FRS, FRAS, hon FIMA." *Bulletin of the Institute for Mathematics and Its Applications,* 12 (1976): 87–90.

Godfrey, C., and A. W. Siddons. *Modern Geometry.* Cambridge: Cambridge University Press, 1908.

McMurran, S. L., and J. J. Tattersall. "The mathematical collaboration of M. L. Cartwright and J. E. Littlewood." *American Mathematical Monthly,* 103 (1996): 833–45.

———. "Cartwright and Littlewood on Van der Pol's equation." *Contemporary Mathematics,* 208 (1997): 265–76.

Nine

Joe Gallian

Deanna Haunsperger

As President of the Mathematical Association of America (MAA), Joe Gallian was a very busy man. Not that he wasn't busy before he became president: The man seems to have only one speed—busy. After three days of meetings of the Executive Committee at MAA Headquarters in Washington in May 2007, on his sixth math-related trip already that year, Joe agreed to talk with me about his activities and thoughts on mathematics one afternoon. I asked if he could spare an hour and a half. Three and a half hours later, he was still going strong, regaling me with personal history, thoughts on the MAA, and wonderful stories. I didn't want the conversation to end.

High School

I grew up in Pennsylvania, in a place called New Kensington, about twenty miles north of Pittsburgh, along the Allegheny River. My mother was a waitress, and my father a foreman in a factory, a glass factory. I have one brother, two years younger.

Figure 9.1 Joe Gallian (right) with his brother and parents in 1952.

My teachers noticed that I was good at math, and I enjoyed it, but I failed history just because I was goofing off. I was a class clown. I didn't intend to go to college. I was born in 1942, and it wasn't all that common when I was growing up. Where I was from, most people didn't go to college; and since no one in my family had gone, I didn't think about going. I only tried to do well in math because I liked it.

Figure 9.2 Gallian's school photo from 1956.

I think I had a very good education in math, probably better than most high school kids now. My math teachers were outstanding. This may sound weird, but people from my generation had the great advantage that there weren't many opportunities for women except teaching and nursing. They didn't become lawyers or doctors, for instance. That meant that people like me had the benefit of women who might have joined those professions becoming teachers. My guess is that the best college math students now do not become high school teachers in the same percentage they used to. Also, back then a lot of men wanted to go into teaching. People with a bachelor's degree in math graduating now, in 2007, have a lot more opportunities than those graduating in 1957 or 1959.

In high school I just hung out with kids in the neighborhood. Although, come to think of it, my three best friends in high school actually became math teachers. I didn't think of that until this moment. Two went to Slippery Rock, which, it turns out, is why I went to Slippery Rock.

The Glass Factory

My father was hired by a factory when he graduated from high school at age seventeen. He worked there his whole life; he died before he retired. In those days, there was no affirmative action. You did not have to give people equal opportunity in hiring. In fact, it was considered a perk for employees that their sons (and maybe their daughters in the office) would be hired in the plant. This was just routine. I don't think you could get a job there without your father or an uncle working there. So I knew that I had this job waiting, and it was a union job. This may sound ridiculous, but union jobs in 1959 paid $2 an hour, and on that you could actually buy a car (not right away of course) and eventually buy a house and have a family. And your wife didn't even have to work. Of course, the standard of living was lower. Each person didn't feel he needed his own car. You could have a very modest home. You know, we had no TV and no telephone, but we didn't think that was particularly odd either. People lived much more modestly. They didn't eat out much, didn't travel much, didn't vacation much; they lived more simply. Nevertheless, that was considered a good-paying job when I got it.

I started out as what they call—(Maybe this is too much detail; when you talk to me, it's like pulling your thumb out of a dike.)—a "peeper." I'm sure it's much more sophisticated now, but the machines had these little peepholes right where the glass is in this molten pool, glowing hot, maybe 4,000 degrees, because it's just melted sand. A metal strip with spikes goes down in this pool, then pulls it up, just like taffy, about four stories. When it goes up, it's very thick, and they put these rollers against it, adjusting the weight to determine the final thickness, ¼ inch, ⅜ inch, or whatever. The peeper's job was to look in this machine [as Joe launches himself off the couch on which he was sitting, and shows me how he would peep into the viewing windows on the glass machines] and see if everything looked okay. I had three machines, first I'd go to this one, then the next one and the next one, making this circle all night long. It wasn't dangerous, but it wasn't very pleasant either. First of all, it was incredibly hot, boiling hot, and secondly, it was incredibly boring. You couldn't talk to anyone; you just went from machine to machine to machine. If you dawdled or did anything you weren't sup-posed to do, someone would shout, "Watch those machines!"

If you thought you saw something, you would clang this gong, and an expert would come and look in, and try to fix it. You had a very narrow window. What would happen is that the glass would start to pull unevenly, and you could see something was start-ing to go wrong. Or there might be a little impurity, a marble-sized lump or something. If you saw this lump, someone would run over and raise the rollers; since if they hit the lump, they'd snap back, and then there'd be big trouble. If the machine went down, it would take about eight hours to bring it back up again. There would be a big mess, and it would be very costly. If a peeper caught one mistake in his whole shift, or even one in a whole week, he would save them more than they were paying him.

"Breaker" was the most dangerous job, but I didn't have to do it my first three years. It was a glass factory, and the glass would be pulled up several stories from the molten sand below, cooling somewhat as it went. A machine would make a little scratch on the still-red-hot glass, then the "breaker" would grab it like this [Joe jumps up once again to show me the technique, arms wide apart clasping an imaginary pane of molten glass taller than he is], and lean back, and it would just break off. They usually made them in very big pieces, and cut them to size later. It was a difficult and dangerous job. After it broke, you put it on a rack, and by that time it was time to get another one.

I worked in the glass factory for three years. It's a really good story how I quit. There were about 1,100 employees, but about 1,200 in the summer. When work-ers took their vacations, someone would be hired to take their place, then be laid off again as time went on. I would work for three or four months, then get laid off; you'd always work and get laid off, work and get laid off. Because my dad was in the fac-tory all those years—his entire life—he had enough pull that he could keep me from having to break. Maybe twenty or thirty guys were hired the day I was, but I was at the top of the list. When they would call you back, you'd have to take whatever job was open, but there were always enough guys called back that I didn't have to be a breaker. Well, one day, after I had worked there three years, a guy said to me, "Next week you're going to break." A few people did like it because it paid extra, and it was a macho job. Also, there was a choice of machine. There were nine machines running, and workers with seniority would pick the best ones. If you actually accumulated some seniority, it might not be that bad. You could get bonuses, and you could avoid the danger. But a lot of guys got injured: cuts, lost kneecaps, big gashes. So anyway, the guy said to me, "You're breaking next week," and I thought, "Oh, no!"

An Eyeball Might Be Worth $5,000

What happened was that you got a three-day break-in period, and that actually wasn't too bad because they put you with an old guy (that might mean a five-year breaker or a three-year breaker), someone with experience who showed you how to do it. But if somebody had experience, that meant he had a better machine since he had more seniority. You broke in on an easy machine, but then had to work on one of the worst machines. So I thought, "Well, this isn't all that bad." Then on the fourth day, I came in, and I was at the very bottom of the seniority list and was assigned the last machine available. Sure enough, this machine hadn't been running well; they'd been having lots of trouble with it. Oh brother! So now I'm there, breaking on my own, and I only lasted about three hours. I was terrified I was going to get cut. You lean way back, and sometimes the sheet of glass would explode; instead of being in one piece, it would shatter, and you're left holding one chunk, and another chunk is like a guillotine. So I called the foreman over and said, "I'm leaving—I'm going home." He said,

"You mean you're not even finishing your shift?"; and I said, "I'm going home." He said, "If you walk off, you lose your job. Do you realize this?" And I said, "I'm going home and"—here's the exact quote—"I'm not a married guy; I want all my parts. I don't want to lose any fingers or kneecaps—I want all my parts!" It was kind of morbid. Say someone lost a kneecap; the first thing he would do is figure out how much he'd get. You got paid if you got injured; an eyeball might be worth $5,000, a finger $1,500. They had a formula. I said, "I don't want any $5,000 or $1,500. I want all my parts."

How 'Bout if I Go to College?

I was on a four-to-twelve shift, and I got home about eight o'clock—I was supposed to be working until midnight. My dad had been working in this factory his whole life, and I come home four hours early! He said, "What are you doing home?" I said, "I quit." And he said, "What do you mean you quit? I don't get you . . . what do you mean, you quit?" I said, "I quit. They had me breaking, and it was so dangerous, I thought I was going to get badly hurt." He said, "I can't get you a job anywhere else; I've worked only one place. We don't have any connections anywhere else. What are you going to do? You're twenty years old, and you're living at home." I said, "Well," and I hadn't thought of this ahead of time—it wasn't as though this had been in the back of my mind—but I said, "How 'bout if I go to college?"

I just threw it out there. And the main reason I threw it out was that he was very upset, and my dad was usually a very low-key guy. And he said, "College? Where would you go to college?" I said, "How about Slippery Rock?", which only came to mind because of those guys I was in high school with. Of course, they were all seniors by then because they had been in college the three years I had been in the factory. I said, "Ray Bitar and John Ciesielski went to Slippery Rock." He said, "What would you do?" and I said, "I'd become a math teacher." His attitude changed completely, and he said, "Well, okay, we'll go up there tomorrow and check this out." Then suddenly he made a 180-degree turn. Sure enough, he called the factory the next day and said he couldn't come in, and we drove up to Slippery Rock, about fifty miles away. We didn't call in advance or anything; we just went up there. My dad said he'd like to talk to someone about his son going to college. When I told the admissions guy my story, he said, "Did you take your SATs?" I said, "No, because I didn't intend to go to college." "What were your grades?" "Very poor. I had As in math, but Cs and Ds in almost everything else." I had to get a B in English in my senior year in order to graduate, my grades were that low. I told him that I graduated near the bottom of my class, that I was just going for passing since all I needed was a diploma. I told him the story about the breakers, and he said, "We weren't planning to take any more students, and you should have taken these exams, but we're going to let you in because we've found that people who go to the military, or work in a factory, and then come here are among our best students." They let me in on probation.

Slippery Rock

Slippery Rock didn't have any dorm space available. However, I found a place just off campus. A woman who worked in the cafeteria and her mother had the downstairs of a house, and they rented out three places upstairs. Two of my roommates were senior math majors, and this turned out to work to my advantage because they weren't very good at math. On the other hand, I was in precalculus, which was not very exciting. They were taking modern algebra from McCoy's book. That starts with rings, and so they were talking about rings and units all the time. They were struggling with it, but I thought that that was far more interesting than my course. I went out and bought a modern algebra book, and started reading it just for fun. After about two weeks, I was thinking, "This stuff isn't that hard," and "These exercises aren't that tough." I ended up tutoring my roommates, covering the modern algebra right along with them.

Figure 9.3 Charlene and Joe Gallian at a party celebrating his college graduation in May 1966.

I Just Studied, Studied, Studied

My strategy in high school had been to do almost no studying, just pay attention in class. When I first started in college, I think I misjudged that it was a fairly big step. After a C or D on my first math exam, I started putting in lots of time. When I got to college, I didn't know how to study, math or anything else. But I learned. Let's say I got a B in biology as a freshman. If I had taken the same course as a sophomore, I would have gotten an A with about a third the effort. As a freshman, I studied the wrong things: I didn't know how or what to study. It took me about a semester to learn study habits and the self-discipline that was necessary. After that I did very well. It turned out that the more I studied, the more I enjoyed it. I had a girlfriend at home, so I didn't date on campus, I didn't hang out with the guys or go to football games or social events; I just studied, studied, studied.

Marriage

I married Charlene at the end of my junior year at Slippery Rock. I was her high school sweetheart. She was a senior when we met, but I had been out of school for two years. Our homes were about ten miles apart, but in different towns. There was a place called Henry's, a little diner with a dance floor and a juke box where you could buy a hamburger and a Coke. It was a hangout for high school kids, and I went there five or six nights a

Figure 9.4 Joe and Charlene Gallian at Turkey Run State Park in Indiana, 1970.

week. Charlene hung out there with five other girls (we called them the "Big 6"), and I would come with maybe one or two other guys. Every now and then one of my friends would ask one of the girls to dance, and another friend would ask another one, and so I asked the one who is now my wife to dance. Because she was still in high school, she had to go home early. One night her mother came to pick her up, and her mother was impressed that I walked Charlene out to the car; that's how the whole thing started. We went together for about three years before we married.

Charlene had a pretty good job as a secretary at a steel company in Pittsburgh and was making enough to support me. When we got married, since I was at Slippery Rock, she began working for the state of Pennsylvania, a civil service job on campus with a decent income. When I went to Kansas, she had a federal government job, working for the University of Kansas ROTC. She was the breadwinner for many years.

Graduate School

When I started at Slippery Rock, a teacher was the only thing you could be; there was no other degree possible there. Then in my sophomore year, they started to offer a new degree, a liberal arts degree. I talked to a professor I thought the world of, Dr. Anthony Pagano. He was very, very stingy on grades, but a great teacher, a really great teacher. I loved the way he taught. He and I became friends, and I'd occasionally talk to him outside class, so I talked to him about switching degree programs. I told him that I wanted to do what he did, I wanted to be like him. He was the image of what I hoped to become. He replied, "Well, then I recommend you switch over to this liberal arts degree," which of course I did.

I applied to a handful of graduate schools; I remember that I applied to Kansas, Minnesota, Michigan State, Wisconsin, Chicago, and Purdue. I was given advice that big state universities had a lot of financial support, and there were lots of options. I had no idea what to do; I was just taking advice from others. I ended up deciding to go to Minnesota because their financial offer was the best. So I wrote the other schools declining their offers, and I wrote Minnesota accepting theirs. This was about a week in advance of the deadline. The day before the deadline, Pagano knocked on my door and he said: "G. Baley Price called,"—he was once President of the MAA—"he's from Kansas,

and he wants to talk to you." So I went up to Pagano's office and called Price back, and he said, "I called you about this letter you sent, and I just wanted you to know that we can offer you this NASA Graduate Training Fellowship. We were supposed to offer it to you in the first round, but there was some kind of mistake." Now, I'm quite sure that what happened was that they had offered it to somebody who then turned it down, but in any case, he said, "If we were to offer it to you, would you accept it?" And I did. I hadn't started college until I was 20, and then I got married while in college, so I wanted to be on a fast track. I thought a fellowship meant no TA-ship, and no TA-ing meant that I could take an extra course or two each semester, and that's exactly what I did.

When I got to Kansas, I wasn't placed in any graduate courses right away, but instead into what were called "senior/graduate level," although most of the students were undergraduates. For abstract algebra, we used Herstein, a big step up from Slippery Rock. Some of the best students were freshmen and sophomores, and a lot of them became professional mathematicians.

When I got into that abstract algebra class, I thought Lee Sonneborn was the worst teacher ever, because I was used to Pagano. Pagano would go down to the finest detail, explaining every step. Sonneborn was just the opposite; he would say, "Well, you can finish the rest yourself," or "This proof is by induction." Or he'd say, "For the proof, mod out by the center." Anyway, he just threw out these hints, and I'd think, "What is this? This guy's ridiculous; that isn't a proof." But his idea was that he'd give you the big steps, and you'd fill in the details. After the first week, I thought he was great. I really admired him, and I had him for five courses in one nine-month period. I knew several of his PhD students who were in this group theory seminar, and I was convinced I was going to do a thesis under him too.

Then Sonneborn came in one day and said that he had taken a job at Michigan State. I was very upset, but there was a young guy named Dick Phillips who was also a group theorist. They both did infinite group theory, so I switched over to Phillips. I took his courses, and he said that I had to write a Master's thesis, which I did with him. But then one day he came in and said that he was leaving for Michigan State as well. Now there weren't any group theorists left at Kansas. By this time I knew I wanted to work in group theory, so I decided to apply elsewhere.

Notre Dame

I ended up going to Notre Dame because they had someone in infinite group theory and they gave me the best offer by far. After I was there about two weeks, I went to the professor who did research in infinite group theory and told him that I had come to Notre Dame to work with him. He responded by telling me that he was leaving! When I told my tale to the department head, he said, "I really think you should consider working in finite groups instead."

I had already signed up for a course in finite group theory since I had never had one. The professor was Warren Wong, who was both a good group theorist and an excellent teacher. He was involved in the classification of finite simple groups, which was an incredibly hot area in those days. Eventually I went to him and said, "I'm looking for a thesis advisor. I'm taking my candidacy exams next week, and I think I'll probably do OK." He replied, "Well, I'm going to New Zealand next year on sabbatical; would you like to come along and work with me there?" My wife didn't want to leave, and I didn't want to either, so I turned him down.

Wong had been a student of Richard Brauer at Harvard. Brauer was one of the most famous people in group representations, a pioneer in modular representations of groups over fields. In a sense you could say he started the classification of finite simple groups, through something called the Brauer-Fowler theorem. It was the first approach to this whole classification problem, although he himself never thought it could be done. I ended up writing my thesis under Karl Kronstein, another student of Brauer's. That worked out very well because he gave me a good problem. That is, I was able to get two papers out of it, published in very good journals. He gave me almost no help beyond giving me the problem, but that worked to my advantage, because when I eventually got to Duluth, there was nobody to work with. But I was already used to working on my own.

Getting a Job

I was at Notre Dame for three years, graduating in 1971. However, I didn't get a job that year, for quite a good reason. About five years ago, the AMS had a chart at the joint meetings, showing unemployment rates for mathematicians. The worst year ever was 1971. When you're at the beginning of a bad period, you don't know that that's the case. In fact, people who graduated in '68 and '69 got great offers. So I wasn't worried, my advisor wasn't worried; there was no hint that things were going to go south. Also there was no affirmative action, which is very important to my story, because you didn't know where the jobs were, they weren't posted. It was literally an old boys' network: If a school had an opening, the department head—say he got his degree at Ann Arbor—would call there and say, "Got any good graduates coming out? We have an opening." That was the way it worked.

When I applied for jobs, Charlene and I both wanted to live where our families lived, so I only applied to colleges in western Pennsylvania and a few places in Ohio. My letters simply said, "I'm going to finish my PhD at Notre Dame. I'm in finite group theory and have submitted a paper. If you have a job for which I might be qualified and you would like more information, please let me know." Of course, at that time there was no word processing or even photocopying. Every letter had to be typed, one by one. Out of whatever it was, twenty-some schools, it might have been that none of them even had an opening. Anyway, I got no replies, nobody asked for an interview, nobody even asked for any further information. So I went into the office of my chair, O. Timothy O'Meara, and said that I hadn't received any responses. He said, "Don't worry about it, you can stay here as a postdoc, but next year

Figure 9.5 Gallian with his sons William and Ronald, 1976.

I want you to look all over." So the next year I sent out 145 applications, and I did get six requests for more information. I sent them the information they asked for, and then I waited. Nothing—I heard nothing from any of them. Well, that's not quite true. Southern Oregon asked me to come for an interview, but then called again and said, "The job's been frozen. There is no job." Finally, I started making phone calls. The other places I called, including Duluth, said that they had hired somebody else. So now I'm down to zero, and it's May. I talked to O'Meara again, and he said, "Don't worry about it, you can stay here another year." I'm extremely loyal to Notre Dame, because they always treated me extremely well from the day I got there. They treat their students and faculty like family. That may sound like a cliché, but there's truth to it.

Oh, I forgot to tell you another part of the story. We adopted two boys during my last year of grad school.

Becoming a Parent— "The Beatles Spoke to Me."

Charlene had always wanted to have kids but wanted to wait until I was ready to graduate. She wanted to have one child of her own and adopt one. You have to remember, this was the sixties. The Beatles spoke to me: "All You Need is Love," "Give Peace a Chance," "Imagine." That's my Bible. That's how I frame my life, the way I think about things. So we thought it would benefit the world if we adopted somebody who needed a family.

When we went to an adoption agency, we of course had in mind some brand-spanking-new baby, perfectly healthy, and so on, like in a movie from the forties. But of course, there are many kids who are hard to place, and those were the ones they showed us. There were two natural brothers, one five and one three, and they were hard to place because they were older, and the agency didn't want to split them up. They showed us their pictures and said, "We can arrange a visit." The boys came once for a short time one Saturday afternoon, and then came back a week later and stayed overnight, and then—I might have it wrong, it was thirty-five years ago—then they came back for another overnight visit. Then they never left. [Along with their sons William and Ronald,

Figure 9.6 The Gallians with their children William, baby Kristin, and Ronald, August 1976.

Joe and his wife Charlene have a daughter Kristin, born in 1976.]

Starting in Duluth

Back to the job story: the important thing was that O'Meara said I could come back. Now it's June of 1972, and it turns out that the person hired at Duluth backed out. When they called to ask if I was still interested in the job I had applied for, I said, "I already accepted a job here at Notre Dame, but they are just doing me a favor. I think that if I had a job offer, they'd get by without me." So I went on that interview.

Hundreds of things came up at the interview, little conversations, but one question I didn't consider important at the time turned out to be crucial: "Every math major has to do a senior project. It could be an expository

Figure 9.7 Gallian is a popular and enthusiastic teacher.

history of a math paper or a report on a paper or a chapter in a book." I was asked, "Are there any group theory topics for math majors to do as a senior project?" I said, "Finite group theory is loaded with them; there are all kinds of good ideas." That question turned out to be crucial because all of the math majors had to find their own project advisors. The other guy who interviewed for the job was an infinite group theorist. When they asked the same question of him, he said, "Group theory's too complicated." He didn't realize it at the time, but it sunk him. When I got to Duluth, I immediately latched onto student projects and had more students than anyone else the first year. And that turned out to lead to my Research Experiences for Undergraduate (REU) program.

The REU Business

A few of my students actually did some original research. I was able to get something in *Mathematics Magazine* based on some of the senior projects, and I wrote a joint paper with a few students. Even in projects that weren't publishable, I tried for some new things, not just exercises. I was trying to stretch the students. In 1976 I read in the *Notices of the American Mathematical Society* (*AMS*) that there was a program called an Undergraduate Research Participation (URP) program—run out of the education division of the National Science Foundation (NSF). It said that there was money available for faculty members to run a summer program in which students would read seminal papers and make presentations to other students and their advisors, and for that they'd get a stipend. These weren't quite REUs because the expectations were lower; it was more like a literature search. I applied for this grant, and sure enough it came in. There were only a handful in the whole United States: I knew of five total, including my own. The great thing is that I had the whole country to myself, in a way. I sent letters to places like Princeton, Harvard, and MIT, inviting their students to apply, just as I do now. Since there weren't many places offering these opportunities, I had really good students from the outset. There wasn't much competition.

Over the years I've had 152 REU students. That doesn't count multiplicity; some were there for two or three years, and they count as just one. I like to take some who are very young, like Melanie Wood. I took her after her first year. I knew Melanie was extra special, first woman ever on the United States team for the International Mathematical Olympiad. I wrote her a letter when she was in high school, and I said, "I hear a lot about you. People tell me you did well in the Olympiad training program. As soon as you get into college I'd like you to write me a

letter asking to be in my program." This she did, and it's been great; she's come every year since. This year she's going to be the coach of the Girls in the Mathematical Olympiad team in China, so I'll get her to come to my program for only ten days, but I'll take her for any amount of time she can spare.

About 150 papers have come out of my REU programs. For about twenty years there have been about the same number of students in the program. Not everyone gets a paper out of the experience, but a fair number write more than one. Melody Chan wrote three, and Jason Fulman got three in one summer, all published in good journals. Doug Jungreis got five one summer and three the next. And by the way, almost all of these are in the *Journal of Combinatorial Theory*, *Journal of Algebra*, *Pacific Journal of Mathematics*, *Journal of Number Theory*, *Electronic Journal of Combinatorics*, and *Discrete Mathematics*. I treat the summer program like a mini-PhD thesis. I want to give them a PhD-level problem, and while they might not do as much as there would be in a thesis, I'm hoping it will be a good start, a basis for a PhD thesis. That's my goal.

At first, my students worked only in finite group theory, because that was the only thing I knew, but then I backed into graph theory by way of Cayley graphs. These are graphs on groups. One referee's report on a group theory paper said, "You are really looking at this the wrong way. These are Cayley graphs." Once I saw that, I thought, "This is really a good thing. This is easy to visualize, and there are natural graph theory questions that might not seem like natural group theory questions, and this is a good way to give students problems." We hit a gold mine. One incredible stroke of luck concerned Hamiltonian circuits on Cayley graphs. This was a rich area that undergraduates could easily be funneled into. Some of those papers are cited very often now, as they actually contain some of the major theorems in the area. David Witte [now known as David Morris] has what I think are the best theorems on the topic, the Witte theorems.

It's a little tricky saying what success means in an REU. Some years ago I had a really great student, nearly perfect in every way: great attitude, hard-working, easy-going, background from Michigan, and so on, but I couldn't find the right problem for him. I gave him one problem that didn't work out; it wasn't his mathematical ability, it was just a poor problem. And then a second problem and a third problem, and by the end of the program, he really didn't have anything to show for it. I couldn't say, "Write up your stuff and send it in." There was nothing to write. I felt really bad about that. About six months later, I needed to get hold of him. I had his home phone number, so I called his mother and said, "This is Joe Gallian; I'm trying to reach your son. Can you give me his phone number?" And she said, "Professor Gallian, it's such a pleasure to talk to you. My son had such a great time in your program. He just enjoyed that summer so much, talking math day and night with all those bright kids. It was just a wonderful summer, and he learned so much." Anyway, I learned a lesson from that: you can still get a lot out of an REU experience even if you don't have anything on paper to show for it. Ideally, I like for students to have a good time, to

Figure 9.8 Gallian with his URP students, August 1980.

network, to get a paper in a respectable journal, and to give a talk at the Joint Mathematics Meetings in January. That's the perfect ending. But even if they learn something, some new branch of mathematics, some new problem in mathematics, if they go through the process, the research process, and learn something, that's a success.

Stories about REU Students

Manjul Bhargava is an interesting story. He agreed to come to my REU, but he had very close ties to a grandfather in India. There was to be a family reunion in India the first week of the program, and he asked if he could go there before coming to Duluth. I said, "Sure, fine." He also said that if I gave him a problem in advance he'd work on it. When he arrived in Duluth, he had completely finished it except for the write-up, so he asked whether I had another problem for him. I gave him a number theory problem, and it turned out that that was beautiful, it worked out so well, because he ended up with a spectacular, completely novel approach. It was in number theory, but he had this ingenious ring-theoretic approach. He won a Morgan prize for that.

When Melanie Wood came, she was a young woman, having just finished her first year of college, and I really underestimated her. I gave her a bad problem, a really bad problem. It was a graph theory problem, and after about a week, she said to me (this is what's very nice about Melanie: it's not that she's impolite, but she's willing to express her opinion), "This is really a bad problem." I said, "OK, but the problem is that you don't have much time. You've already lost several weeks from arriving late to the REU, and now you've lost another week on this bad problem. You've only finished one year of college, and I don't really have any problems that would be good for you. I said, "Manjul has a lot of open problems, but I'm afraid they're too hard for you." She said, "Let me take a look at them." She came in the very next day and said, "I love this." It went really well. She wrote a very nice paper that got accepted by the *Journal of Number Theory*, a very high-class journal. In large part, she won a Morgan Prize for that.

One of the youngest people ever in my program was Danny Goldstein. He started as an undergraduate at the University of Chicago at sixteen. When he went to the Olympiad training program, he was the youngest person ever to be invited at that time. When he got into my REU program, he was eighteen. He had already taken graduate algebra at Chicago. The summer he was in Duluth I was working on the first edition of my abstract algebra book. I don't include Galois theory when I teach this course at Duluth; it's too difficult. Nevertheless I have a chapter on Galois theory in the book for others. There were some exercises on Galois theory in the book, which meant that at one time I could do them quickly. However, when I tried to do one of them again, I couldn't recall how I had solved it before. So I called down to the student apartments. That year there were three students in the program, Danny Goldstein, Larry Penn, and Irwin Jungreis. Irwin had finished sixth on the Putnam while Danny and Larry were in the top five, so they kept joking all summer long about how Irwin was the slacker, the weak one of the bunch. Anyhow, I called down there; it didn't really matter who answered the phone, but Danny answered and I said, "Danny, I have this Galois theory problem, and I can't figure out how I solved it. Can you solve it?" He said he'd call me back. Fifteen seconds later he called and said, "Here's how you solve it."

In 1994 the AMS and MAA had a joint summer meeting in Minneapolis. I don't normally take my REU students to the summer meeting, but since this was in Minneapolis, the travel cost would be minimal; we'd just drive down. Ron Graham was President of the AMS at the time, so I wrote him a letter and said, "Ron, I'm bringing my whole REU down there, and I'd like to invite you, in your capacity as President of the AMS, to meet these students and have either breakfast, lunch, or dinner with us." Ron has always been supportive of my program, and I got this nice reply, "Yes, let's make it dinner." As soon as I got that letter, I wrote to

Don Kreider, President of the MAA, "I just invited Ron Graham for breakfast, lunch, or dinner, and I'd like to do the same for you to give the MAA equal opportunity." So on one day, these students had breakfast with the President of the MAA and dinner with the President of the AMS.

Beatles Forever Completely Changed Things for Me

A great thing about having been born in 1942 is that I was 13 in 1955 when rock and roll was born—Elvis, Little Richard, Fats Domino. I listened to the Pittsburgh radio stations that would play rock and roll. My father loved music, but it was a completely different kind. In those days, there was just one record player, no TV, and maybe two radios. I listened to my dad's records, and he'd have to listen to mine. My dad had lots of records, Bing Crosby, Al Jolson, the Ink Spots, Tony Bennett, Frank Sinatra. I would hear those types of music all the time, and also they played them on the radio all the time. Then when rock and roll came, I started buying rock and roll records. By that time I had a little attic room. It wasn't finished, but it had electricity, and I'd play my rock and roll up there. I liked all my dad's music, but I thought this new music was really energetic and great stuff. I thought that rock was exciting, and I bought lots of records, and I had the words memorized.

Then some time around 1959, the original wave went down. Elvis was drafted into the military. Buddy Holly died in a plane crash. Little Richard got religion. That early era just

sort of died. Then they had Frankie Avalon, Bobby Vinton, Fabian, these artificial, manufactured singers—they didn't have the raw edge. They didn't have the energy and the personal magnetism that I thought the early people had. They'd try to get someone who looked good. Little Richard wasn't handsome. Buddy Holly wasn't good looking. Elvis had sex appeal, but these other people didn't. I hated their music. I sort of lost interest in rock and roll and turned to folk music, Bob Dylan and Joan Baez. When the Beatles arrived—February 7, 1964—it took me a month or two to get to like them, but when I did, I just loved them.

Now I have a tremendous collection of Beatles stuff. There's a museum in Duluth called the Depot. They have a summer exhibit, and one summer the whole exhibit was my stuff and that of another Beatles collector in Duluth. Big billboards in town said, "Beatles Memorabilia Exhibit at the Depot." I've spent many, many, many thousands of dollars on Beatles stuff, and I still buy tremendous amounts. There are a thousand Beatles books in print, and I probably own about three or four hundred. What's great about having a hobby like the Beatles is that there's always new stuff. Always. It's unbelievable. So every birthday, every holiday, every Father's Day, my family can buy me Beatles memorabilia. I probably have more than a dozen Beatles neckties.

I listened to the Beatles for years and years, and then one day, as a Christmas present, my wife brought me a book called *Beatles Forever,* and it completely changed things for me. This is not an exaggeration; it changed my life. I hadn't realized that you could study the Beatles in an academic way. In this book, the author carefully analyzed, the way a sociologist would study some tribe or something like that, the Beatles. Many people have written dissertations on the Beatles, and how they're impacting the culture. For about twenty-five years now, I've regularly taught a course on the Beatles. In fact, I now teach it fall, winter, and spring. I first started teaching the course through the sociology department. The curriculum committee, which had to approve the course, asked what my qualifications were for teaching a humanities course. I wrote back and said, "I've spent more time studying the Beatles than I did writing my PhD thesis." That's true—by now it's about 20:1. And they accepted that. It's an incredibly popular course. Not an easy one, though.

Project NExT

In 1998 when Jim Leitzel died, and Chris Stevens needed a replacement, I became a director of Project NExT, an MAA program for new PhD graduates who are about to enter the teaching profession. Chris's strategy was to get two people to replace Jim Leitzel. I'd already been helping Project NExT and so had Aparna Higgins. We had both been presenters, so it was a natural fit. That too changed my life. It's very likely that without it, I wouldn't be President of the MAA; my visibility in Project NExT gave me a big advantage.

Project NExT energizes the profession with all the young people contributing. For example, a lot of math journals hunt for

Figure 9.9 Aparna Higgins with Gallian, enjoying another successful year of Project NExT.

referees. I can put a message out to the Project NExT fellows that the *College Mathematics Journal* and *Mathematics Magazine* would be happy to have Project NExT fellows serve as referees. We always get outstanding responses. The MAA has about 130 committees, and we often look for Project NExT volunteers to serve on these committees. They energize the profession in that way. I didn't go to a national meeting and give a talk in my first years after my PhD. We tell them to go and give talks; this is what they should be doing. Current Project NExT people think this is perfectly natural because we tell them it's natural. I've got this mnemonic, CEO. We give them the Confidence, the Encouragement, and the

Opportunity to make a contribution to the profession quickly.

The networking is incredible among the NExT fellows. If you looked at the network of mathematicians I had after five years in the profession, theirs is often greater by the end of one year. One way that this happens is through the colored dots; people get a colored dot on their name tags for every Project NExT meeting they attend. It's an automatic opening for a conversation in an elevator, at a talk, or wherever. NExT really has made a difference.

I give the NExTers a regular, one-hour spiel of advice. There's a phrase, a catch phrase, that they always throw back at me. "Just say yes." I got it from Nancy Reagan, because she said to "Just say 'no' to drugs." I say, "Just say yes." If somebody asks you to be on a committee, just say "yes." If somebody asks you to teach a course that you've never taught before, just say "yes." If somebody asks you to come give a talk, just say "yes."

When you join a new department, your top priority is to make yourself valuable. That's why I say, for example, if they need somebody to teach a new course and no one else wants to do it, you can volunteer. That's my advice. A huge number of people have told me they've followed my advice. Not 100 percent of the time, of course, but they try to keep it in mind, and they say, "You know, it does work."

I tell them there are many ways to succeed and many ways to contribute. One of the ways to contribute is through the MAA. The MAA is a friendly organization, especially to younger people. You're going to be welcomed. The MAA is interested in how you can contribute. Can you serve on a committee? Are you willing to do work? I say, make yourself known, go to a Section meeting and say, "I'd like to help out; do you have anything I can do?" I tell them to write to the editors of journals and say, "I'd like to referee for your journal. Here are my credentials: I got my PhD in finite group theory, I wrote the following three papers, and I'd like to start out as a referee." My advice is to immerse yourself in the profession. Teach, do service, do some research, volunteer. The more you do, the more you learn; the more opportunities you create, the more you succeed.

Heroes

Two of my heroes are Ron Graham and John Conway. Why them? Well, first of all, they're incredibly versatile mathematicians. Secondly, do you realize how often they say "yes"? For example, Ron Graham spoke at my sixty-fifth birthday conference. John Conway spoke at the Hudson Valley Conference for Undergraduate Research some years ago. Not many mathematicians at his level would be willing to take the time to speak to a group of students, but Conway did. Conway and Graham give talks at MAA meetings. Ron Graham took the time to serve as President of the MAA. In addition to being outstanding mathematicians and being incredibly versatile too—they take the time to be part of the math community, and I admire them for that.

I have other heroes. Lennon and McCartney are my cultural heroes. Martin Luther King, Robert Kennedy, Paul Wellstone,

Ralph Nader before the year 2000, those are my political heroes. Some people might say, "You know many of your heroes were assassinated." It's not that they're my heroes because they were assassinated, they were assassinated because of what they stood for. The fact that they stood for something admirable is why I admire them so much.

President of the MAA

When I came to Duluth my chair, Sylvan Burgstahler, was very high on the MAA, and he was definitely encouraging people to join. I started going to meetings and giving talks, and that way I met people. It was a very positive experience, but I didn't join at the time. My pay was low, and I just didn't want to pay the dues. In my fourth year there, I said I wanted to be considered for promotion. Bill McEwen, the division chair and a member of the math department, asked to see my *vitae*. He said, "I noticed you aren't a member of the AMS or the MAA. Why's that?" And I said that I didn't see any particular benefit to it. He said—and this has stuck with me ever since—"Whether it has a direct benefit to you or not, it's your duty to support your professional organization." My wife and I contribute to various organizations (like Mothers Against Drunk Driving) even though we don't directly benefit from them. It's the same philosophy. It's an obligation that one has to the larger community. Once I joined, the opportunity arose to edit the section newsletter. I agreed to do it, and from that became President of the North Central Section.

One of the MAA's strengths is our volunteerism. The MAA is very welcoming; I like the camaraderie of it. We can easily take in a lot of energy and talent and benefit from it; young people can make a contribution quickly. If the MAA weren't around, it would have to be invented. There really is a niche for us. Although the AMS does a good job communicating to the general public, I think we are needed there, too. Certainly our work with students, like the math competitions and the Putnam exam, is definitely a major contribution to the profession. Without such opportunities for students, there's a good possibility that Melanie Wood and others like her would not be mathematicians.

It would be nice if the MAA had a great source of money with which to work. The Halmos gift of $3,000,000 for the restoration of the Carriage House has made an enormous impact, of course. That sum going to the AMS would not be such a big deal, but to us it's a major step forward. As president, I wanted to find additional support for Project NExT, and I think I've actually done quite well on that. I also want to improve the public awareness of mathematics, its value and its beauty. I am happy with what I have done with that, too. We have a documentary film about the 2006 U.S. Math Olympiad team coming out in 2008, we are working on a women-in-math poster, and we have an exciting public lecture series at the MAA headquarters.

The MAA has given me a tremendous amount. It's an incredible honor to be president. The REU has been unbelievable. I couldn't have imagined in 1972 when I arrived in Duluth, at a branch campus in northern Minnesota, that I would get the chance to work with some of the top math students in the country. It's amazing that this could have happened, such a series of serendipities in which one good thing led to another. It's a good example of how you can make things happen. You can chart your own course. If you're hustling, saying yes, doing things, trying to help others, you can end up doing really well. I never could have predicted such a thing.

You're Breaking Tomorrow

MP: As our interview drew to a close, and I was packing up my three-and-a-half hours of tapes, Joe sat back and became quietly reflective. He had just given me his life story to this point in a nutshell; an impressive, active, joyful life spent in service to mathematics and the mathematical community. I could tell his eyes were seeing a different place, a different time. "My brother still lives in our hometown where I grew up. He knows everyone in the town and what they're doing. I can't even remember my neighbor's name. We're just so different." Realizing that that could have been him, except for a few turning points in his life, Joe jerks upright again, breaking the tranquility with his explosive laughter, "And it's all because of that one phrase: 'You're breaking tomorrow!'"

Ten

Richard K. Guy

Donald J. Albers and Gerald L Alexanderson

As a young faculty member at Goldsmiths College in London, Guy was told by a seasoned staff member, "Goldsmiths is noted for its successes and suicides—which are you going to be?" A half-century later the answer is clear—Richard K. Guy is a great success. He has published more than two hundred papers and nine books, some running to several volumes and some translated into languages other than English. And more books are on the way. His writing is marked by clarity and wit. He is a very popular lecturer, who has an unerring sense of when to drop the unexpected pun on his listeners. His punning is usually accompanied by a sly smile and subtle twitching of his notable eyebrows. Guy recalls his first day of kindergarten. Three-year old Richard told his parents, who were very eager to know how the day had gone, "It was all right, but the teacher doesn't know much. She asked me what shape the world was and all sort of things that I thought she would have known." Guy is a collector of unsolved problems and has published two books devoted to them. He always seems to be working on other books and several papers. His prodigious production is perhaps explained by the fact that he works both when awake and when asleep. He thinks, in fact, that most of his serious mathematics is done while sleeping! In his spare moments he climbs mountains, camps in the snow and even on glaciers, and composes chess endgames.

MP: *You have covered a lot of the world in your professional career, but I understand that you started life in England.*

Guy: Yes, I was born in Nuneaton, Warwickshire. But let me go back even further, 1912, when my father met my mother. She was the headmistress of a big girls' school in Acocks Green in the south of Birmingham. He was a young school teacher, eight and a half years her junior. He decided that he'd go out to the colonies, as I suppose they called them in those days. He went off in 1912 to teach in Perth Boys' School in Western Australia, the idea being that my mother would eventually join him and they would be married. In that case I would have been born Australian. But World War I started, and my father, swept along by the wave of nationalism, joined up along with many colleagues. He was just in time to be at the fateful landing at Gallipoli, one of the few who survived. It must have had a very traumatic effect on him. He also served in France for a while and was wounded slightly. So he came home and married my mother. They never made the second attempt to emigrate. During the last five years of his life my father came out to stay with us in Calgary.

MP: *Did your parents have any particular mathematical talent?*

Guy: No, but they were general elementary school teachers in the old tradition so that they learnt and taught all subjects, including mathematics. I don't really know what my mother's specialty was. She was just an expert on everything. My father claimed that English was his first subject, and he could certainly quote great chunks of Shakespeare. He also taught crafts, woodwork I think we would call it. He was very good at crafts. They were both very good craftspersons. You can't think of a craft—metal work or basket work or bookbinding or whatever—that they did not do, and with a very high standard. This was inhibiting for me because, although I knew how to do all these things in theory, as soon as I started to practice them, I would either give up or they would take over with "No, no, you should do it like this." My mother particularly was a great perfectionist.

MP: *Although your parents were both school teachers, they discouraged you from a school teaching career. You were the first of how many children?*

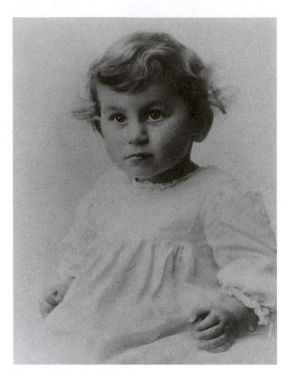

Figure 10.1 Baby Richard

Guy: I was the first of one. That's it: an awful, spoiled only child. Of course there are enormous advantages and disadvantages to being an only child, but it also means you grow up to be a rather selfish person and you aren't always able to accommodate to other people around you later. You get very possessive and don't want people invading your territory. My mother had an incredibly strong personality. She was one of those infuriating people who are always right, even when they're clearly wrong. My father just abdicated all responsibility, even in financial matters. I think he would have been irresponsible as far as money is concerned. School teachers were very poorly paid, but my mother was a financial wizard as well as everything else.

Figure 10.2 Guy and his parents with their "baby" Austin car.

She managed so that we had a very good standard of living for those days. Teachers in Britain took a 10 percent cut in salary during the Depression, and salaries were small enough before that. So she managed extremely well. We were even able to run a motor car from the year 1924. She used to buy a baby Austin car, sell it at a loss of about £15 at the end of two years and buy a new one. So we always had a car, which was very unusual in our social stratum.

The First Rule of Good Health

MP: *Pleasant?*

Guy: She was too demanding of high standards in everything to be pleasant, but she did have a good sense of humor and was far from being unpleasant. I'm incredibly proud and glad to have had her as my mother. I was lucky with parents. The first of good health rules is to choose your parents carefully. And I made an extremely good choice. Also they both had very good principles: they were always impeccably honest, straightforward, and outspoken against anything that was not for the common good. Mother was a great leader in the Women's Institutes, for example. She founded several Women's Institutes in Warwickshire, where I grew up.

MP: *What were Women's Institutes?*

Guy: They were really to cultivate the traditional image of women and could be formed only in villages, in small communities. They taught housekeeping and country arts, but they also taught crafts. They would have meetings and someone would demonstrate how to cut out a pattern for a dress, for example. I can remember helping my mother make the diagrams that she would hold up to show the women how to take measurements.

MP: *How then did you choose something as purely theoretical as mathematics?*

Guy: I didn't choose mathematics. It was chosen for me.

MP: *By whom?*

Guy: That's a long story. I was obviously fairly bright as a child and showed some signs of being a prodigy at about three in that I learnt tables and did mental calculations well beyond my years. During school I certainly disliked history, geography, English, and most languages. I think I would have been good at languages. When I went to Warwick School, which was where I got the main part of my education, I was in fact at the top in French. But I'd had a year of French and most of my peers hadn't, so it was not too difficult to be at the top. My interest was killed off by having for three or four years in succession a master who was very unsympathetic.

The subjects I was really interested in were mathematics, physics, and chemistry. When you get to the sixth form in the British system, which is where you start specializing, you settle down to three subjects. Then it was clear that those were my three subjects. At that time if you'd asked me which was my true subject (I was very young—I was only thirteen when I went into the sixth form,

which is two to three years younger than usual—I would have said chemistry. I was very keen on chemistry. But the chemistry teacher was a gentleman in the old English sense of the word. I think he had a private income. He didn't need to teach at all; he just did it out of love of the subject. So he had a very detached attitude, saying, "Here is my subject. If you care to sit at my feet and listen, you are very welcome." And that was fine with me. I would have gone on lapping it up; however, the mathematics teacher was a real go-getter and livewire, and if he saw anyone around who showed any aptitude at all he grabbed him and said, "Now, come on, you're going to do mathematics." He grabbed me and said that I was to do mathematics. That was Cyril T. Lear Caton, Lear because I think he was related to Edward Lear. He was a very interesting person. He was only about five feet tall—his wife was even shorter, I think. He was a very notable math teacher, and these were the days in Britain when all the senior math teachers in the so-called public schools, the grammar schools, were first class honors people from the two older universities. He went off to be headmaster at Alcester Grammar School after a while. After he left we got Kenneth Lansdell Wardle, who not too long ago retired from Birmingham University. Again he was a very good mathematician. I learned all sorts of things, such as Lagrange multipliers. I could solve any differential equation that you could solve with the usual methods. Quaternions I knew about, and therefore vectors in particular— wooden vectors, as I like to call them.

MP: *Wooden vectors?*

Chess, Bridge, and Snooker

Guy: Well, 3-D vectors, the kind I still tend to be fascinated with because you really can see them. Much of my career I've spent teaching engineers. I enjoy this very much because you can see the mathematics working in the ordinary three-dimensional space that we think we live in. It helps you to keep your feet on the ground. Anyway, Wardle then took over for about three years. So when it came time to decide on a career, obviously I'd just go on doing mathematics. When I went up to Cambridge, I went to Caius College, which was where Wardle himself had been. At the end of my time there I still didn't know what I was going to do. In those days one or two people went into the actuarial profession, and one or two people stayed on to do research mathematics. I got only a second class degree because I spent most of my undergraduate life playing chess or bridge or snooker. So the only thing I could do was to go into school teaching. My parents had both strongly advised against their profession, because it was so underpaid, but there was nothing else. In those days one of the deadly sins was to be unemployed. Your curriculum vitae had to show continuous employment, and if you'd ever been out of work no one would give you a job. I made only a few ineffectual attempts to get a job. Another difficulty in those days was that you had to have had experience. How anybody without experience manages to get experience is always a bit of a difficulty and still remains so to this day! I didn't succeed in getting a job, and in order to avoid this terrible hiatus, I went to

Figure 10.3 Guy as a member of the Royal Air Force in 1942.

Birmingham University and took a teaching diploma, really to fill in and give myself another year in which to make a more serious attack on landing a job, which I eventually did.

I went to Stockport Grammar School, which boasted among its former pupils W. L. Edge, the geometer, who died recently, and Horace Lamb who wrote the classic book on hydrodynamics. So I had landed at a place with a very good mathematical environment. But within a year or so I got overtaken by World War II. When I came back, apart from having gained a wanderlust as a result of traveling during the war, it became obvious to me that although the school was very good from the mathematical point of view, if I really wanted to advance my career, I

Figure 10.4 Officer Training Field Day at Warwick School.

couldn't do it by staying there. Eventually, if you stay, you become the senior mathematics master, but there were two excellent teachers of mathematics who had been at the school already for twenty years, and it was clear that they were going to be there for the next twenty. I was just going to be teaching middle school mathematics, and enjoyable as that was the first few times around, I could see that it was going to be increasingly boring. So I decided I'd better try to move on. I applied to Goldsmiths College, a part of London University but a teachers' training college. From the academic point of view I don't think anybody regarded it as being on the same level as other colleges of the University.

MP: *I notice that you had a first class honors on the Tripos.*

Guy: Well that was Part I. That's put down just to impress people. I only got a second in Part II, which is called Senior Optime and also sounds quite impressive (first class is called Wrangler).

Master's Degrees for Five Guineas

MP: *But then in 1941 you took the MA at Cambridge.*

Guy: Well, again, that was completely phony. Let me put it in the best possible way that I can. The mathematical education at Cambridge was so superb that getting an ordinary bachelor's degree there was equivalent to getting a master's degree anywhere else. So for the price of five guineas and a decent interval of time (three years) you automatically got a master's degree.

MP: *Does that mean you applied for a master's degree?*

Guy: Yes, you have to know the Cambridge system. The masters become voting members of the university. After you acquire a certain amount of seniority you're allowed to take part in the political process. It's not really an academic matter; it's a university political thing.

MP: *Did this require any additional time at Cambridge?*

Guy: No. Only the length of time to write out the cheque and make the application.

MP: *What's the theory behind that system?*

Guy: You'd have to ask a university historian. It probably dates from the time when everybody got a bachelor's. You just went on residing there. If you had been there long enough you were sufficiently senior to be admitted to the brotherhood. You were elevated to a kind of peerage and allowed to take part in the discussion of college and university business.

"I Don't Regard Myself as a Mathematician"

MP: *Had you solved any interesting problems at this time?*

Guy: Your title, *Mathematical People*, did reassure me when I agreed to be interviewed. I was glad that the word mathematician doesn't appear in it, because I don't regard myself as a mathematician.

MP: *You don't? A person who's published nearly 200 papers!*

Guy: Again, you can fool some of the people all of the time, and all of the people some of the time. And you publish or perish; this is the way you survive. I have published virtually nothing in the way of theorems. If I sent you a list of publications, you would find that there are few things which you might call research papers. Otherwise it's all padding. When I came to this continent, I found that you had to have a publications list, so I started dragging things together. OK. My first theorem is a very nice one. If you look in an early issue of the *Mathematical Gazette*, roughly the British equivalent of the *Monthly*, you'll find "A Single Scale Nomogram." I merely made the observation that a cubic equation with no x^2 term has zero for the sum of its roots. If you draw a cubic curve, $y = x^3 + ax + b$ and put a straight line $y = mx + c$ across it, the sum of the x-coordinates of the intersections is zero. If the curve is symmetrical about the origin ($b = 0$) and you change the sign of x on the negative half, then one coordinate is equal to the sum or difference of the other two. Combine this with the principle of the slide rule, which simply adds one chunk to another. For example, if the chunks are logs, you have multiplication and division. Anything you can do with a slide rule you can do with any

Figure 10.5 Guy in 1938.

single-scale nomogram. That was my first theorem, I suppose.

 MP: *You said with some passion a minute ago that you're not a mathematician. If you're not a mathematician, then what are you?*

Guy: An amateur, I mean I'm not a *professional* mathematician. I'm an amateur in the more genuine sense of the word in that I love mathematics and I would like everybody in the world to like mathematics.

MP: *Why should they?*

Guy: Well, because it has become a very important part of our lives, yet the great majority of people obviously have a fear and hatred of mathematics. This is a great shame. It's almost like going back to very

superstitious times when many people were bedeviled by the idea that there were evil spirits lurking behind every tree, and lived in fear and trembling. I think the modern equivalent is people living in fear of mathematics when it can be such an enjoyable thing, and such a useful thing. My desire has been to pursue mathematics, mainly in the selfish way of just enjoying it on my own, but also wanting to pass this enjoyment on to other people, particularly as I get older and feel that at least I owe something for the terrific privilege that I've had of being able to live, all the time doing what I wanted to do. Most people have to earn their living doing things they don't like doing, and I've always been amazed that people would pay me to do what I would be doing anyway. Now I want to repay people by trying to sell mathematics to as many other people as I can.

Figure 10.6 The Guy family at Rhonegletscher in 1949: (L to R) Mike, Anne, Peter, Louise, and Richard.

Figure 10.7 The Guy boys at Cambridge: (L to R) Kenneth, Peter, Richard, Andrew, and Mike.

MP: *You say you're an amateur, but I don't know if you want to stick by this statement that you're not a mathematician. How about your collaborators—Berlekamp, Conway, Erdős? Are they mathematicians?*

Guy: Certainly they are. And in fact this rather supports my remark that I'm not a mathematician. I appear to be hanging on to the coattails of some of these more obviously genuine mathematicians. I love mathematics so much, and I love anybody who can do it well, so I just like to hang on and try to copy them as best I can, even though I'm not really in their league.

MP: *Do you seek them out?*

Guy: Well, you would have to take them individually to get an answer to that question. I've had the advantage of age over most of them. I've noticed that young people are much more respectful of age than they were say one generation ago. Two or three

generations ago you had to be respectful of your seniors or else. But then somewhere about the fifties, perhaps even earlier, youth got very rebellious against age. I notice nowadays that young people don't seem to mind the generation gap. They're willing to listen and talk to older people.

Conway was still a young graduate student when I met him through my son Mike, who is two or three years his junior. So if Conway didn't have respect for me as a mathematician, he at least had some respect on account of my age. Also I think that my son had sold him on some of the things that I'm interested in, particularly game theory. Mike Guy was interested in games because what one would call my second theorem, I suppose, was the discovery of the periodicity of what are now known as nim values, what we called $\mathcal{G}$-values in the paper with Cedric Smith. So when I met him, Conway had already been a bit brainwashed by Mike in the areas of game theory, graph theory, the beginnings of combinatorics, and so on. The real reason Conway would bother with me at all is that he had enormous respect for Mike Guy. Mike is a genuine mathematician, although you'll find only about one paper of his in the literature because he doesn't bother to publish. In Conway's writings, scattered about, you'll find references to him. In "Monstrous Moonshine" Mike Guy appears early on in the paper. And Conway's paper on the Archimedean solids in four dimensions was really a piece of joint work with Mike. Conway has enormous respect for Mike's abilities and therefore a certain respect, perhaps vicariously, for the parent that produced Mike Guy.

The Benefits of "Hmmm, Yes"

Guy: I think Conway liked to use me, also, for what John Leech calls a blind ear. You know how very useful it is when you're just on the verge of getting some mathematical idea, to have someone who will listen. Our daughter used to use my wife in this way. My wife doesn't know any mathematics at all, but when my daughter would be wrestling with some topological problem, she'd come downstairs from where she was studying. Louise would be in the kitchen, baking a cake or something, and my daughter would start expounding away. Louise would say, "Hmm, yes." If you've just got somebody there listening, you're clarifying your own ideas. I think Conway used to do this with me and found it useful to have someone say, "Hmm, yes," every now and again. Anyway for one reason or another, a very unlikely pair of people got

together. Also I knew him in the early days when he was completely unable to write anything down. He was pouring out ideas all the time, so I would scribble them down and say, "Look, you've got a lovely paper. Here it is." I can't say I wrote his papers for him—that's certainly not true—but I often gave him a draft which would then provide sufficient stimulus for him to put the thing down the way it really should have been written out in the first place.

With Elwyn Berlekamp, again, it was perhaps more the respect for age than anything else, but the respect came out of game theory again. I first met Berlekamp in 1966 at the Chapel Hill Combinatorics Conference. He was rather pleased with himself, as he deserved to be because he's extremely bright. He had just solved the children's game of Dots-and-Boxes. And an important tool in this was the Guy-Smith paper I mentioned earlier. He'd used this as part of the quite complicated theory that we regard as being one of the most satisfactory chapters in our book *Winning Ways*. The general public would find it rather heavy going, but to a mathematician I think it's a good chapter. Berlekamp said, "More people should know about the Guy-Smith paper. It's a very important one." He said we should write a book on games. I said, "Yes, that's fine with me." I thought about it a bit and said, "Well, I know someone who could help us a great deal with this." So I wrote to Conway. Once a year I would go back to Cambridge for a few weeks, and we'd sit down and Conway would spout away, and I would scribble down things. So a draft of a chapter would be produced.

MP: *By you.*

Guy: Well, you often see these papers produced by great mathematicians that just say notes were prepared by some graduate student. Much of Conway's writing has originated in this way. I would write things down as he dictated them and try to write out a fair copy. The next time I'd see him I'd show it to him and he'd say, "No, no, you've got this all wrong."

MP: *The gestation period for* Winning Ways *was fairly long.*

Guy: It was started in 1966 and published in 1982, sixteen years later. But we were not working continuously. We are three very different personalities, and I think my main role was keeping Berlekamp and Conway both together and apart. A lot of the work was done in our house in Calgary. Louise has been an enormous help to mathematics and to a large number of mathematicians, including some rather queer ones, whom she has given hospitality to at one time or another. Occasionally Conway and Berlekamp would almost come to blows. It's very interesting that they can work together because the lay mathematician would guess that Conway and Berlekamp were very different kinds of mathematicians. Once Conway wrote a letter of reference for Berlekamp for the chair of mathematics or electrical engineering or whatever it is that he holds now in Berkeley. Conway just dictated it to me. He said how well he had worked with Berlekamp because Berlekamp had these great intuitive ideas. His own analytical abilities could cash in on them.

Figure 10.8 Guy, Conway, and Berlekamp holding a copy of *Winning Ways* (Photo by Klaus Peters).

Now I think most people would guess that Conway was the intuitive one, with brilliant ideas, and that Berlekamp was more the steady plodder. But he isn't. Having worked with them both I know that Conway is not a plodder, but he works fantastically hard and methodically and pounds away at something until he gets there. Berlekamp gets quite bright ideas, suddenly. He just states theorems, and they're true. He doesn't stop to prove them. Conway's much more cautious. He says, "Wait a minute. How do you prove this?" And Berlekamp says, "Well, you know, it's quite easy." Often there were wrangles that would go on and on, and it would turn out that there were holes in Berlekamp's ideas. But they could always be mended, so that it was a very interesting collaboration.

The book *Winning Ways* is a collection of essays. The ideas came from Berlekamp and Conway, but they were mainly written by me, almost entirely from Conway's dictation, either of his original ideas or of the ideas written out by Berlekamp in what was for us a not very acceptable style of writing—not acceptable because Berlekamp tended to a rather formal mathematical style, which we certainly didn't want for *Winning Ways*. We wanted it to be a popular book. Conway would get hold of Berlekamp's rather long and ponderous essays on various topics and tear them apart and say, "Let's invent some names." We would delight in doing this, lots of plays on words. We would try to outdo one another in producing the most outrageous puns.

MP: *You did very well.*

Guy: I remember one occasion when the two of us visited Bell Labs when Berlekamp was still there. Every now and again Conway and I would be working away together and something would set us off. I can remember at least two occasions when Berlekamp almost lost his temper. He wanted to get on with the work, and we were making puns back and forth or going off into some complete irrelevance, nothing to do with the mathematics. It was like slapstick comedy. But it was wasting time as far as he was concerned.

MP: *Yes, but it was a very effective collaboration.*

Guy: Oh, yes.

MP: *You've worked in a variety of fields: number theory, combinatorics, graph theory, geometry, and game theory.*

Guy: Jack of all trades and master of none.

MP: *At the same time you've had some very distinguished co-authors, of course, a whole string of them. Did you pick the co-authors for their fields or the fields for the people you could work with?*

Guy: Well, my own interests have been in combinatorics, game theory, and number theory. I seize hold of people who are in those fields. Erdős collaborates with everybody. It's hardly fair to say anything positive or negative about someone because he has written a paper with Erdős. It's such an enormous crew and includes some of the world's best mathematicians, but it also includes young hopefuls who have written only one paper in their lives because Erdős stimulated them. Erdős is another person who had a big effect on me. I first met him at the Davenport seminar in 1949. I can't claim that he had very much influence on me then, but when he came to Singapore in 1960, he stayed with me and gave me two or three of his problems. I made some progress in each of them. This gave me encouragement, and I began to think of myself as possibly being something of a research mathematician, which I hadn't done before.

Dickson—Better Than Shakespeare

MP: *In a conversation with Erdős some years ago—actually it was in* Mathematical People— *he said that he no longer works much in number theory because all the problems that are left are much too hard. Do you think that is valid?*

Guy: Yes, it is, I think. Number theory is at least four thousand years old, because we know the Sumerians must have been doing number theory to write down tables of what we now call Pythagorean triples and approximations to the square root of 2 and things which are clearly mathematics and not astrology or astronomy or anything else. So number theory is the oldest branch of mathematics, and quite a lot of people have put in a lot of thought. If you want to get into an area, then number theory is perhaps the field where it is hardest to get up to the frontiers of knowledge. Combinatorics is an extremely young subject, and it has in common with number theory an immediate appeal to amateurs such as myself in that it's very easy to understand the problems. But it may be of varying difficulty to solve them. I was always interested in number theory, ever since I was a small child. When I was a youth I'd bought L. E. Dickson's *History of the Theory of Numbers*.

MP: *How old were you?*

Guy: Well, seventeen perhaps. I walked into a book shop: an ordinary book shop but it had a copy of Dickson's *History*, and I was fascinated, although it was about six guineas, which was a fantastic price in those days for me. I wasn't earning anything, so I must have talked my parents into giving me the money. But it was better than getting the whole works of Shakespeare and heaven knows what else.

MP: *Let's go back to the field of combinatorics for a moment. Unhappily, it seems to me that although we've had absolutely first-class work in the last ten, fifteen, twenty years in combinatorics with the work of L. Lovász, E. Szemerédi, and so on, there is yet to be a Fields Medal awarded in this area. It seems that the subject is not recognized within the mathematical community at large. The mathematical establishment seems to be dominated by algebraic topologists, algebraic geometers, and such.*

Guy: There's not much to say. You've already said it. There are fashions in these things, and combinatorics has not yet been very fashionable. Of course, you can go back and point to a lot of Euler's work, and some of Gauss's, and many people, like Cayley and Sylvester. A lot of what they did was combinatorics. Even Hardy, Ramanujan, and, I'm sure, F. G. Frobenius worked on combinatorial problems. There are plenty of good names. You can say, "Well, that's really a combinatorial theorem." But combinatorics hasn't on the one hand formed itself into a coherent body of knowledge, and on the other hand it hasn't caught the attention of the serious mathematicians.

MP: *Yes, I suppose the algebraic geometers and the topologists would look to their fields and say, "Well, it has this enormous structure." This may not be true of combinatorics.*

Guy: I think this is the trouble. Many people regard combinatorics, probably quite correctly, as being just a bag of more or less unrelated facts, and a bag of particular tricks for solving particular problems. You can pick out a number of themes that do carry over into various problems, but, as you say, it's difficult to see a real structure there.

Figure 10.9 Louise and Richard Guy on Heart Mountain (Photo by Steven Bryan Grantham).

MP: *It isn't a lack of depth.*

Guy: Pólya's theorem (the Redfield-Pólya-deBruijn-Read theorem) and Ramsey theory are really very deep and of enormous significance. You can do a great deal with them.

MP: *I want to get back to Erdős for a minute. I know your Erdős number is considerably less than one.*

Guy: How do you get an Erdős number of less than one? Do you divide by the number of joint papers?

MP: *That's exactly right. So what is your current Erdős number?*

Guy: I had a letter from the Hungarian Academy the other day. Somebody there is trying to put together a list of all of Erdős's papers. So they wrote and said, "We notice the following paper or papers joint with Erdős. Could you check the bibliographic details and tell us of any other papers that you know about?" So I did make a list. But there were only four papers with Erdős. So I have Erdős number ¼.

A Distinction in Art

MP: *You often illustrate your manuscripts. As a child were you interested in art?*

Guy: Yes, as I've already said, my parents were school teachers, and every summer they went away to a summer school for teachers, refresher courses of many kinds, which tended in the main to be moderately light-hearted. Many were in art and crafts at which both my parents were superb performers. When I was very small I used to go and play on the sands with my bucket and spade. When I came to be a youth, I enrolled in an art course, or a so-called art course. But I can't in any way claim to be an artist. When I was at school, I took a subject which went by the rather grandiose title of "Engineering" but in fact was half metalwork and half machine-drawing. We drew stress diagrams, which was very useful to me because it was really mathematics and statics. I did machine drawing and developed a certain amount of facility. I won't claim to be in the Escher class. Escher was in the first instance not so much an artist, although I'm running him down considerably in suggesting that, but he was an incredibly meticulous designer. He was a very careful draftsman in the more technical sense. You see the same thing in Salvador Dalí's early work. Not only do you get the strange visions and surrealism of Dalí, but you also get this terrific technical ability. So that even in the early days when Dalí was a complete shocker, as far as most people were concerned, a large number of people had to be respectful of his technical ability. I had acquired a certain amount of technical ability as I went to these art courses, and I learned lettering. That's a skill that has been very useful and that I passed on to my daughter and to my granddaughter. It was something that I was fascinated with—designing fonts and typefaces and posters. I took a teaching diploma at Birmingham University. I needed some nontechnical subjects, so I took art. I got a distinction in art then because it was very easy to turn out this mechanical kind of thing. You didn't have to be an artist in any sense or show any originality. You just had to show some sense of taste and a certain amount of ability and technical skill. I've always had that, and it's always been very useful to me.

How to Write a Preface

MP: *I want to turn to the preface of* Winning Ways.

Guy: Yes, guess which of the three of us wrote it. I always thought that everyone would guess that it's Conway.

MP: *Oh no, I guessed it was you.*

Guy: You're quite right.

MP: *Let me read a little bit here to you. "Does a book need a preface? What more, after fifteen years of toil, do three talented authors have to add? We can reassure the bookstore browser, yes, this is just the book you want. We can direct you if you want to know quickly what's in the book to the last page of this preliminary material. This in turn directs you to pages 1, 255, 427, and 695. We can supply the reviewer faced with the task of ploughing through nearly a thousand information-packed pages with some pithy criticisms by indicating the horns of the polylemma the book finds itself on. It is not an encyclopedia. It is encyclopedic, but there are still too many games missing for it to claim to be complete. So just don't stand back and admire it, work of art though it is. It is not a graduate text since it's too expensive and contains far more than any graduate student can be expected to learn. But it does carry you to the frontiers of research in combinatorial game theory and . . . many unsolved problems." That's an absolutely delightful preface.*

Guy: So much so that I know at least two reviewers who just copied that paragraph. Eventually after working all these years, Berlekamp more or less worships the ground that Conway and I write on. He has really learnt style. We even taught him how to spell. He has an enormous respect for what Conway and I have managed to put together after buffooning around. Conway more or less said, "Anything that you write will be OK with me." In any case, I tried to write the preface as though Conway was writing it. By that time I'd assimilated his style.

Mathematics, Games, and Geometry

MP: *I am afraid that at this point we have gotten away from the chronology of your career, which has taken you to a couple of rather exotic places.*

Guy: Unfortunately most of the changes in my life have been made for purely negative reasons. I've already told you that I moved from Stockport Grammar School to London, and not because of anything I disliked about the school. It was a lovely situation. I used to help with the coaching of cricket and rugby. I enjoyed teaching young people. I started a chess club there. In fact, let me interrupt myself to say something which I think is important. It may throw some light on the curious working of what I'll call my mind. When I first started teaching at Stockport Grammar School, I was given a Form, that is, a class of say twenty-eight boys. You're sort of moral tutor for this Form, in addition to being a mathematics teacher. Mine was Form 1, the year before they started doing four years of algebra and geometry, leading to what used to be called School Certificate and is now O-level. The only mathematics they did was arithmetic: fractions, decimals, percentages, and so on, for the sixth year in succession. Most of them already knew the stuff backwards, and we had six periods a week because we worked Monday through Saturday. I felt that I would get terribly bored even if they wouldn't. So I declared that we

Figure 10.10 Ninety-year-old Guy with nine-year-old Kaavya Jayram in 2008, who is already writing research papers in number theory, also one of his research areas.

would in fact have four periods a week of arithmetic, and the two other periods would be called Games and Geometry. "Geometry" consisted mostly of geometrical constructions and manufacture of polyhedra, and "Games" was mostly chess. We also used to play Battleships, that would introduce coordinates and a certain amount of strategy. But I was already interested in chess endings. I would demonstrate to them how certain of the more combinatorial aspects of the endings went: for example, if you are left with rook and king against king, you mate the king. There are similar endings of that kind which are done mechanically. If you really want to be a competent chess player these things should be among your tools. So I taught these in a fairly formal fashion. When I came back to the school after the war, the chess club was functioning every night, and they had to press nearly all the staff in the school into supervising it. There was a boy who had just left the school and was already third in the British championship, Alan Phillips. It was very much like what they were doing in Russia. You taught chess in the schools and produced an enormous body of very good chess players.

The geometry was also rather fascinating. I can tell you a nice Mordell story arising out of this. I said, "For homework this week make any polyhedron. Make a pyramid or something." There were a couple of kids who came with an icosahedron and a dodecahedron which I certainly hadn't expected. I was hoping to lead up to those, of course, but we'd done only cubes and tetrahedra. They came with them welded in metal. So I said, "Well that's very, very good, but where did

you get the mathematical expertise and the metal working expertise to produce these? Who helped you with them?" And they said, "Oh, our father has a big collection of polyhedra." And I said, "Oh yes, what does your father do?" They said, "He's professor of zoology at Manchester University." It was Herbert Graham Cannon; he was quite a famous figure. And so I got an invitation to tea one Sunday and got to see Cannon's quite notable collection of polyhedra. He just had it as a hobby. So I got to talking, and of course he was in the same Common Room as Mordell. I knew of Mordell, of course. In fact I'd written and asked him some of the stupid questions that I often get asked myself these days, and, quite reasonably, he hadn't bothered to answer. Mordell was one of many mathematicians who were interested in climbing. Cannon told me this lovely story about Mordell, how he'd arrived back on Monday morning in the Common Room and proudly announced to the general multitude, "Well, twenty-four hours ago I was climbing." This went over like a lead balloon. Except for the English professor, who looked up from his *Times* and asked, "Trees?" Louis tended to look a bit anthropoid.

Singapore

Anyway, I left Stockport because I couldn't see any future as far as advancing my career there was concerned. Then when I got to Goldsmiths—and this was a very exciting place to be—I worked much too hard. I spent a lot of time on chess endings, a lot on graduate mathematics, a lot on the job, and far too little on my family. I was teaching twenty-two periods a week, and many of the lectures were pretty much university level courses. It was a small place with about five hundred students, but they used to run two rugby teams, two soccer teams, two field hockey teams, and put on plays, and do everything that you can think of that students used to do, an incredible amount of activity. It was a fascinating place to be, but the academic powers that be suddenly decreed that the general degree, the ordinary degree, was not a suitably distinguished thing for the University of London to offer.

Now they couldn't do very much about the unique position that the University was in. Throughout the colonies and many other parts of the world there were thousands and thousands of people every year who took London degrees externally. Examinations were shipped to people in India and Malaya

and almost every other place in the world which either didn't have very many universities or had very poor ones. So many people managed to acquire this certificate of merit. I suppose in a way it was a substitute for the Open University in the early days. It was an incredibly useful service. So the University Senate couldn't get rid of that, but what they could do was say to Goldsmiths, which was an *internal* London college, "You're not going to teach this stuff any more." With the big guns they had it looked inevitable that we would lose our degree work. And fascinated as I was with the teaching of mathematics, my main interest was in mathematics itself. And there were about half a dozen colleagues in English and other subjects who took off at the same time. We decided, "It's time for the rats to desert the sinking ship." We applied for jobs. As I said, Goldsmiths now is an extremely reputable place, especially in mathematics, but it

Figure 10.11 Guy, front row left, was part of the Staff Cricket Team of Goldsmiths College in 1949.

Figure 10.12 Harry Golenbeck, editor, *The British Chess Magazine*, making a move against Guy in a simultaneous display at Goldsmiths College in 1949.

Figure 10.13 Guy with Teoh Gim Hock working with polyhedra in Singapore in 1952.

looked then as though it might not be. So we took off, and it happened that the first place that offered me a job was the University of Malaya in Singapore.

MP: *That's a long way from England.*

Guy: Yes, but there was the need to survive. I didn't want to go back to school teaching, and to go to anything less than a college didn't seem right to me. The regular universities in Britain could get all the people they wanted in those days, so there was little hope of getting a job there. Going to the colonies was really the only way to get the kind of job I wanted. By this time I had a wife and three children, and I dragged them protesting to Singapore. It's a very curious thing, because when I first met Louise, she was the one who wanted to travel. But as soon as we got married, the nesting instinct came into play, and she wanted to make a home for the children. I'd gone off to the war and traveled

about and got terrific wanderlust. Anyhow, off we went to Singapore. I wouldn't change that for the world. We had an excellent department there, although it was very hard to get people. It was really only a man and boy operation. Well it was two men and a boy because we did have Oppenheim, now Sir Alexander Oppenheim, whom I visited just recently and who's well into his eighties now. We also had Jack Cooke, who was professor of applied mathematics. They were both very reputable mathematicians. I must tell you a story. Bobby Ho, who was in the geography department in Singapore, said to me one day, "Come round to dinner; I want you to meet a friend of mine who's a mathematician." They had both been at King's College, London. The friend was Eric Milner, who had been supervised in a master's degree by Coulson.

The Road to Calgary

MP: *Charles Alfred Coulson? Who was later professor of chemistry in Oxford and very prominent in the movement to unite the churches?*

Guy: Yes, that's the one. Evidently—this is probably an apocryphal story—at a drunken party someone said, "You mathematicians are useless: you couldn't get a job in the real world." As a result someone stuck a pin in *The Times* and said to Eric, "I bet you can't get this job," selecting a job at random. The job was with the Straits Trading Company, a tin-mining operation in Malaya. At the time that Eric came out, this was a pretty dicey business, because of the bandits or terrorists, the Japanese Liberation Army. The day after I met Eric I went to Oppenheim's office and said, "Hey, there's a mathematician lying around loose here in Singapore. Let's grab him." So we did. Well, he's fairly eminent now, a fellow of the Royal Society of Canada, and invited speaker at some of the International Congresses. Later on when I was head of the department I also managed to recruit Peter Lancaster, now another Fellow of the Royal Society of Canada. And both he and Milner are with me now, but for opposite reasons. I'm in Calgary because Peter Lancaster went there first, and Milner is there because I went there. We had excellent people in Singapore, and although we were isolated, we did get people like Heilbronn, Coulson, Eilenberg, Erdős, and the Lehmers to visit.

An important turning point in my life was the Lehmers' visiting Singapore early in 1959

and telling me about the Boulder conference, to which I went. I made contact with dozens of number theorists. It was the beginning of a long friendship with Selfridge. It was the first time I had met Leo Moser. That was another turning point: Leo was one of the various reasons I went to Alberta. I had heard of Alberta through Leo, but I'd never heard of the Calgary Stampede [a rodeo for which Calgary is well-known]. I hope that Leo would be suitably flattered that he was more important than the Calgary Stampede.

Figure 10.14 Leo Moser in Santa Barbara, 1967.

Figure 10.15 John Selfridge, Waterloo, Canada, 1966.

I had met Leo by correspondence, but there I met him in the flesh. We immediately clicked. Leo had this deadpan Buster Keaton sense of humor which many people didn't understand at all. They just thought he was being rude or at least uninteresting. But Leo and I hit it off very much the way John Selfridge and I hit it off.

Do you know Leo's story about this publisher saying, "Well, when's this book of yours going to be published?" And Leo said, "Oh, I don't know, probably posthumously." The publisher said, "Well, make it soon."

"The Numbers Racket"

The Boulder conference was a three-week summer institute in number theory, which is quite a long time, particularly if you get to know someone immediately. A small group of us ended up going around together. Leo Moser, John Selfridge, Raymond Ayoub, and I found ourselves in Trader Vic's in Denver, having a drink. The waitress came up to take our order—she could easily tell we were from out of town—so she asked, "What do you people do?" Right off Leo Moser said, "We're in the numbers racket."

I used to play a lot of bridge and someone wanted to make up a foursome—so I found myself sitting opposite Leo Moser. This was early on in the conference, and I had only just met Leo. He said to me, "Do you know the Smith convention?" I said, "Yes." So we sat down and started playing and after a little while, Leo looked over his hand and said "I've only got twelve cards." I said, "Well, I've

got fourteen," and we threw the hand in. We were not playing for money.

Then there was an interval of around twelve years—it must have been about 1970—and I was staying with Ted and Peggy Youngs in Santa Cruz over Christmas. Some friends of theirs—he was evidently a very well-known film director—were there, and it was again suggested that we have a game of bridge. I had to admit that I knew the rules. I sat down opposite this film director and our opponents were Ted Youngs and his son. The film director said, "Do you play the Smith convention?" I said, "Yes." So we pulled this stunt again. I never saw Youngs get so angry. He became apoplectic. I did know that he had a heart problem, and I got worried that we might have precipitated something. There was a terrific uproar. Eventually it resulted in this film director and his wife leaving the house, and I felt absolutely terrible being part of the sequence of events. But I was reassured that every time this film director came to their house—and they were very old friends—they managed to get sufficiently drunk that they found something to have some fantastic disagreement over. Every time the evening broke up in this way. There was nothing unusual about it at all. So there were two times in my life that I have been able to use the Smith convention. There are, of course, two responses. If one says that he has twelve cards, it means that he has an indifferent hand. So you can respond that you have fourteen, meaning that you too have an indifferent hand. But if you have a good hand, you can say, "I have thirteen; you'd better count again."

Figure 10.16 Guy with Ted Youngs and Gerhard Ringel at Oberwolfach, 1967 (Photo by Michael Kleinert).

As a postscript to this, Patrick Browne, my present head of department, told me that he and Eric Milner were playing a tournament in Calgary, against some university students, when one of the students said, "I've only got twelve cards." Quick as a flash, Eric Milner preempted with "I've got thirteen, you'd better count again." He and Patrick made a small slam!

India

MP: *Between Singapore and Calgary you spent some time in India. What prompted the move to New Delhi?*

Guy: I was reading *The Mathematical Gazette* and to my surprise there was an advertisement in it for a job. In those days, job advertisements didn't appear in *The Mathematical Gazette*. The advertisement was for someone to start up a mathematics department in the new Indian Institute of Technology in Delhi, although it barely mentioned the name. I

don't think it had yet been approved by the Indian government. I thought, "Well, this looks interesting." So I applied. I just wrote a letter saying, "I've just seen this. It's ages old and I presume you've filled the post by now, but if you are still interested, I can send you a more formal application." Back comes a letter by return post, "Yes, please apply immediately." So, I sent it in. Back came a reply again. "Can you be interviewed?" In fact I'd arranged to go on a holiday up in the jungle, and a native arrived with a message in a cleft stick saying, "Could you be in Kuala Lumpur in forty-eight hours' time for an interview?" Anyhow, they were in a mad hurry to get someone. In any event the Indian government failed to approve this for a heck of a long time, so I didn't actually go to India for quite a while after that. I signed a five-year contract, but I did only about three and a half years in Delhi, because although it was a very fascinating and interesting country, the director there had a chip on his shoulder about the British Raj [the time when India was under British rule]. He only knew one word and that was "No." He had excellent people. He'd got really good British people as heads of departments. Engineering in India at that time was the prestigious profession. In most countries it's medicine, but in India it was engineering. So we got the best students from all over India. And we got the best Indian staff. We had help from the British taxpayer who paid our salaries and help from British industry, which supplied the equipment. We really had everything going for us. It was an excellent place, and it had terrific momentum, but we had this director who kept saying "No" to every idea we had,

Figure 10.17 Guy played Archbishop Cramer in *A Rose without a Thorn* in Delhi in 1964.

even if it wouldn't cost money. He just had to be in charge. So several of us didn't finish our terms.

Anyway I was looking for a job again, when Louise wrote to Edna, Peter Lancaster's wife. He had left Singapore in 1962 and had in fact called on us in Delhi on his way to Calgary. Louise wrote to Edna saying, "Richard's pretty frustrated here." So I got a letter back, "Why not come to Calgary?" So I said, "Why not?" I applied to Calgary and said, "I'll come for one year." That was in 1965.

MP: *Quite a shift in climate.*

Guy: It certainly was. But that's fine because I was getting into my late forties by then, and as you get older climates matter more. The Delhi climate is particularly bad. The temperature in Singapore never gets up to blood heat. It's never much below either. But I think once the temperature gets above blood heat, there's nothing you can do about it. That's my criterion for rejecting a climate. If it gets warm you can perhaps run a bath of water and lie in it and hope. If it gets cold, you can run around, or burn the furniture and keep warm. But you can't keep cool.

Endgames and Unsolved Problems

MP: *Spoken like an Englishman. At what point did you start getting interested in all these unsolved problems with which your name is now associated?*

Guy: I think I was always interested, and I kept looking at problems, mainly supplied by Erdős and others. Leo Moser was a great source of unsolved problems. When I found I couldn't solve the problems myself, the only thing to do was to pass them on

Grundy's Game is played with heaps of beans: two players alternately choose a heap and partition it into two unequal heaps, so that the game ends when all heaps are of size one or two: the last player is the winner.

Dawson's Chess is perhaps most easily explained as a bowling game, played with rows of pins. The two players alternately knock down any pin, together with its immediate neighbors, if any, so that a move may separate a row into two shorter rows. Again, the player who knocks down the last pin wins.

These two games are examples of impartial games, in which the options are the same for each player, regardless of whose turn it is to move. The last-player-winning versions of such games are covered by the **Sprague-Grundy theory,** which states that (any position in) any impartial game is equivalent to a heap of beans in the game of Nim. The number of beans in the heap is called the **nim-value** of the position. **Nim** is played with heaps of beans—a move is to select a heap and remove any positive number of beans from it. The elegant theory of Nim was given by C. L. Bouton around the turn of the century.

The *nim-sequence* for a game played with heaps is the sequence of nim-values of heaps of 0, 1, 2, . . . beans. For Nim itself this is obviously 01234567. . . . For Dawson's Chess it begins 01120311033224052233011 . . . and eventually becomes periodic with period 34. Grundy's Game starts 00010210210213213243043 . . . ; Mike Guy has calculated ten million nim-values without finding any periodicity.

to other people. Since I can't do the mathematics myself, I encourage other people to try to do it. In the early days when I first moved to London, when I went to Goldsmiths College, I found myself within easy cycling distance of Thomas Rayner Dawson, the famous chess problemist. Although Dawson was not particularly interested in endgames, he ran the "Endgames" section of the *British Chess Magazine* in addition to the "Problems" section. Because I had been composing endgames and had made a bit of a name for myself, he suggested that I take over the "Endgames" section. He gave me a little section of his card index which referred to endgames, and I eventually built up a very nice endgame library, which I passed on to my successor. Dawson was an amateur mathematician—he has half a dozen papers in *The Mathematical Gazette*—he was a rubber chemist by profession, although he had retired by that time. We used to love to discuss problems. They often turned out to be combinatorial: for example, Dawson wrote on polyominoes long before Sol Golomb did, but he didn't call them polyominoes.

Dawson showed me a problem of his, which we now call "Dawson's Chess." I managed to analyze this and in doing so rediscovered what we now know to be the Sprague-Grundy theory. I was taking a course in number theory from Estermann in University College. I showed him the analysis and asked him if it was significant. He said, "I don't know, but I can tell you who does," and sent me over to the Galton Lab to find Cedric Smith.

Now Cedric Smith turned out to have been an exact contemporary of mine in

Figure 10.18 Cedric Smith playing games, Waterloo, 1966.

Cambridge, though we didn't meet there. He is one quarter of Blanche Descartes, a fictional author who is credited as writing papers in graph theory. They were all undergraduates at Trinity College. B was Bill Tutte, originally a chemist, who developed his expertise in graph theory at that time. L was Leonard Brooks, who became an income tax inspector, but has remained an amateur graph theorist: all graph theorists know Brooks's theorem. A was Arthur Stone, the Syracuse topologist, and C was Cedric. Now they didn't like BLAC, so they changed it to Blanche, and M. Filet de Carte Blanche was born (filet for networks, in which they were interested). Later, as sex change operations became popular, he became Blanche Descartes.

Cedric was quite excited about my rediscovery of Grundy's theory (we didn't know about Sprague then). Grundy had (apart from "Grundy's Game," whose analysis still defeats us) only a few trivial examples where he could apply the theory, whereas I had an infinite family, based on Dawson's Chess,

and Kayles, another game I analyzed. (In Rouse Ball's *Mathematical Recreations and Essays*, there used to be a very significant diagram, due to Michael Goldberg, analyzing Kayles as far as twenty skittles: Goldberg was unlucky not to have discovered the complete analysis and the S-G theory.) This resulted in the Guy-Smith paper, which Berlekamp was to find useful later on.

MP: *Who got you interested in chess?*

Guy: It's hard to say, because neither of my parents played chess, but when I got to boarding school I just picked it up. So I started playing at a moderately early age, nothing prodigious, but I played fairly steadily while I was in school, and we did have a school chess team that played semiserious chess. I used to play for the university but not on the top boards. I suppose I'd always had a combinatorial turn of mind. The endgame, which is capable of exact analysis, appealed to me very much, and I used to love to get hold of books on the endgame. In fact I once planned a large three-volume book on endgames.

Brooks's theorem is concerned with coloring the vertices of a graph. If the maximum valence of a graph is n, then the vertices of the graph may be colored with n colors with no edge having the same color at each end, *except* in two cases: if $n = 2$ and the graph contains an odd cycle; or if $n > 2$ and the graph contains $n + 1$ vertices each of which is joined to the other n.

Doing Mathematics While Sleeping

MP: *How do you work as a mathematician?*

Guy: If I do any mathematics at all I think I do it in my sleep.

MP: *Do you think a lot of mathematicians work that way?*

Guy: I do. Yes. The human brain is a remarkable thing, and we are a long way from understanding how it works. For most mathematical problems, immediate thought and pencil and paper—the usual things one associates with solving mathematical problems—are just totally inadequate. You need to understand the problem, make a few symbols on paper and look at them. Most of us, as opposed to Erdős who would probably give an answer to a problem almost immediately, would then probably have to go off to bed, and, if we're lucky, when we wake up in the morning, we would already have some insight

Figure 10.19 A happy Guy in Toronto, 1967.

into the problem. On those rare occasions when I have such insight, I quite often don't know that I have it, but when I come to work on the problem again, to put pencil to paper, somehow the ideas just seem to click together, and the thing goes through. It is clear to me that my brain must have gone on, in an almost combinatorial way, checking the cases or doing an enormous number of fairly trivial arithmetic computations. It seems to know the way to go. I first noticed this with chess endgames, which are indeed finite combinatorial problems. The first indication that I was interested in combinatorics—I didn't know I had the interest, and I didn't even know there was such a subject as combinatorics—was that I used to compose chess endgames. I would sit up late into the night trying to analyze a position. Eventually I would sink into slumber and wake up in the morning to realize that if I had only moved the pawns over one file the whole thing would have gone through clearly. My brain must have been checking over this finite but moderately large number of possibilities during the night. I think a lot of mathematicians must work that way.

MP: *Have you talked to any other mathematicians about that?*

Guy: No. But in Jacques Hadamard's book on invention in the mathematical field, he quotes some examples there where it is fairly clear that people do that kind of thing.

There was someone earlier this week who was talking about Jean-Paul Serre. He said that if you ask Serre a question he either gives you the answer immediately, or, if he hesitates, and you push him in any way, he will say, "How can I think about the question when I don't know the answer?" I thought that was a lovely remark. At a much lower level, one should think, "What shape should the answer be?" Then your mind can start checking whether you're right and how to find some logical sequence to get you where you want to go.

MP: *Do you think number theorists are different from other mathematicians? Do they think of themselves as different?*

Guy: That's a very difficult question. I think I have to answer that somewhat obliquely. Insofar as I am not a number theorist and I do have some other interests, I would have to claim that they are not different. But insofar as I am a number theorist, I would have to say that we are in a class by ourselves.

Postscript: Professor Guy's concern, expressed on page 177, over the paucity of Fields Medalists recognized for work in combinatorics has been allayed by the awarding of Fields Medals to Timothy Gowers in 1998 and to Terence Tao in 2006 for their work in combinatorics along with contributions to other branches of mathematics.

REFERENCES

For those who would like to investigate further some of the problems and games referred to in the interview, Professor Guy has provided the following set of references:

Berlekamp, E. R., J. H. Conway, and R. K. Guy. *Winning Ways for Your Mathematical Plays.* London: Academic Press, 1982.

Bouton, Charles L. "Nim, a game with a complete mathematical theory." *Annals of Mathematics,* 3, no. 2 (1901–1902): 35–39.

Brooks, R. L. "On colouring the nodes of a network." *Proceedings of the Cambridge Philosophical Society,* 37 (1941): 194–97.

Conway, J. H. "Four-dimensional Archimedean polytopes." *Proceedings of the Colloquium on Convexity, Copenhagen, 1965* (1967): 38–39.

———. *On Numbers and Games,* London and New York: Academic Press, 1976.

———, and S. P. Norton, "Monstrous moonshine." *Bulletin of the London Mathematical Society,* 11 (1979): 308–39.

Dawson, T. R. "Problem 1603, *Fairy Chess Review." The Problemist, Fairy Chess Supplement,* 2, no. 9 (December 1934): 94; solution, ibid., 2 no. 10 (February 1935): 105.

———. "Ornamental squares and triangles." *Mathematical Gazette,* 30 (1946): 19–21.

Descartes, Blanche. "Why are series musical?" *Eureka,* 16 (1953): 18–20; reprinted ibid., 27 (1964): 29–31.

Guy, Richard K. "A single scale nomogram." *Mathematical Gazette,* 33 (1949): 43; 37 (1953): 39.

———, and Cedric A. B. Smith. "The G-values of various games." *Proceedings of the Cambridge Philosophical Society,* 52 (1956): 514–526; *Mathematical Reviews* 18,546a.

Rouse Ball, W. W. *Mathematical Recreations and Essays,* 11th ed., revised by H.S.M. Coxeter. London: Macmillan, 1939, p. 40.

———, and H.S.M. Coxeter. *Mathematical Recreations and Essays,* 12th ed. Toronto: University of Toronto Press, 1974; reprinted New York: Dover, 1987; pp. 36–40.

Eleven

Fern Hunt

Claudia Henrion

If you ask mathematicians why they went into mathematics, you will often hear something like "It was fun; I was good at it; it seemed like a natural thing to do." But for a black woman born in 1948, there was nothing natural about going into mathematics; indeed, not a single black American woman received a PhD in mathematics until 1958. So for Fern Hunt, who later became a professor at Howard University and who is now a senior researcher at the National Institute of Standards and Technology, pursuing mathematics meant beginning a journey through unmapped territory. It was hard work, for as Hunt said, "I was no one's fair-haired boy."

For the most part, Hunt did not receive encouragement from the usual sources: early teachers (with one notable exception), peers, or role models. How, then, did she decide to pursue mathematics, and what enabled her to achieve a successful life as a mathematician?

For Hunt, commitments to research and teaching have been intimately intertwined. Teaching gave her the opportunity to convey to her students that life is much more than "what you own, the car you drive, and where you live." She tried to share the deep

excitement of an intellectual life. At the same time, even with a demanding teaching load at Howard University, it was important to her to continue her research in mathematical biology and applied mathematics. In 1993 Hunt accepted a full-time position at the Applied Computation and Mathematics Division of the National Institute of Standards and Technology. There she works on mathematical problems that arise in research on the physics and chemistry of materials important to U.S. industry. She continues to have contact with undergraduate students by lecturing at colleges and universities and working with summer students.

Fern Hunt's experiences in first predominantly white or mixed schools and then later at a predominantly black university shed light on the advantages and disadvantages of each of these environments.

Growing Up

Fern grew up in a housing project in the middle of Manhattan, not far from Lincoln Center. It was not long after World War II, and there was a great deal of idealism about these housing projects. They were well cared for at the time; trees and shrubs were planted throughout the project, creating a haven in the middle of the city. Her father was a postal employee, and her mother went back to work when Fern was seven, as a transcribing typist for the Welfare Department. Being in a predominantly black environment created a kind of buffer from discrimination.

From the start, Fern was very independent. She was essentially an only child for seven years, until her younger sister was born. She describes herself as "a difficult kid who resented being told what to do. I was a little spoiled, probably . . . and I was not

Figure 11.1 Six-year old Fern after a piano lesson.

a terrifically outgoing person." At first she didn't care much for school and was more absorbed with emotional and personal issues than with doing well in her classes. But at the age of seven, she changed her attitude when she realized that she was in a "slow" reading group. That, she decided, was unacceptable. At the time there was a tracking system in the public schools of New York City, which began as early as second grade.

> It seemed to me that I could do better than this ["slow" reading-group placement]. I was comparing myself with people in the group and already knew more than they did. It just seemed to me that they were going very slowly. So at that point I started making some extra effort. Bit by bit [I worked my way up], jumping two or three groups. I started slightly below normal in reading, and ended up reading a grade or two above by the end of the year.

From this early age, she was well aware of the inequities in the public school system. Although the schools she attended were integrated, the lower-level tracks were predominantly black, and the upper tracks were predominantly white. It was the upper-level classes that had the best teachers and offered the best education. Seeing this, Hunt became motivated to do well both in her classes and on standardized exams in order to ensure that she received the highest-quality education available.

But even getting into these top classes did not erase the different treatment she received because of her skin color. Most of her teachers were white. "This often meant that some of the smaller-minded ones tended not to pay very much attention to you. You didn't get very much encouragement at all." So she made a point of doing so well academically that it was difficult to ignore her. Not until junior high school did she have a teacher whom she liked, a woman mathematics teacher, Freida Denmark, of whom she says, "She treated me with respect. I think she liked me. And that was definitely different." Building up a student's self-esteem was not a focus of most teachers at that time, "and even if they did, people tended to build up the self-esteem of people of the same color." Although Hunt was a bright student, most teachers did not try to encourage or really connect with her.

At this early stage of her life, Hunt was much more interested in subjects she learned on her own than those she learned in school. What first excited her about science was the books she read and a Gilbert chemistry set that her mother gave her when she was recovering from an appendix operation. She was nine years old. "This was a fantastic thing for me because I would just play around and do these experiments. It's kind of strange because if you look at the cover of the chemistry set—I had the junior level—it showed two boys working. I realized that it might be a difficult career [for me], but I thought I'd worry about that later."

A mathematical spark was not lit until much later, when she started learning algebra rather than arithmetic. In elementary school, "mathematics was a subject that I hated. Arithmetic was not very interesting. I always had difficulty with bodies of knowledge made up of arbitrary rules, and that is what math seemed like." But later, when she was thirteen or fourteen and began to develop an aesthetic sense for art and music, she also began to appreciate mathematics.

Algebra—"Things Fit Together"

I began to get more of an aesthetic sense of mathematics when I started doing algebra. I think I did so well because it was a body of knowledge where things fit together. There was some structure, and that fit with my sense and appreciation of structures. It wasn't arbitrary pieces of facts, higgledy-piggledy, but some kind of inherent relationships between facts. That began to change my attitude toward mathematics. But I still didn't see myself as a mathematician because the people around me who were good at math were boys—and white.

Throughout Hunt's early life, the images and messages that surrounded her consistently implied that she would not fit in as a mathematician or a scientist—both because she was female and because she was black. How then, did Hunt come to believe that she *could* be a mathematician?

In ninth grade, Hunt finally had a black teacher, Charles Wilson. That changed her life. He was a science teacher with a master's degree in chemistry from Columbia University.

Hunt: It was really a lucky thing—a chemist with an excellent education and a first-rate mind. He was the existence proof that a career in science for a black person was possible. I badly needed that at that age. He had a wonderful laboratory stocked with all kinds of things, from fetal pigs to electrical motors to Bunsen burners and chromatography tubes. He knew chemistry, physics, and biology, including biochemistry, and would help me and other students with setting up our science fair projects. Mr. Wilson used the Socratic method in his teaching, which was difficult for us at times, but in the end it taught me how to think like a scientist. He was the right person at the right time, and so he had a very profound effect on my life.

MP: *Would he talk to you, also, about going on in science?*

Hunt: Oh, yes. I told him my interests and ambitions. He did two very important things. The first was to tell me about the Saturday program which was available through Columbia University for kids interested in studying science. In fact, this program led to my decision to become a mathematician. The second was that he encouraged me to apply to Bronx High School of Science. If you could point to a single person who had a major influence on me, it would be Charles Wilson.

Bronx High School of Science

Hunt was indeed accepted to the very selective Bronx High School of Science. But to her disappointment, it was not the intellectually stimulating environment she had hoped for. Bronx Science was a large school with three thousand students and a male-to-female ratio of two to one. Most of the students in Hunt's classes were a year or two younger than she was. As she remembers it, the atmosphere was immature and repressive—there was a lock on everything, and monitors were everywhere. Competition was the norm. Although the students were quite bright, most were primarily focused on getting into Ivy League colleges and high-paying professions. Few, she felt, were "genuinely interested in ideas." And the few teachers who made an impression on her were not in mathematics or science but rather in French and history. "The key thing is not so much subject area as much as the kind of teaching approach you take with students who are going to deal with ideas in their adult careers. A teacher should foster curiosity and intellectual energy. That's a rare commodity, and young people who have it need to see examples of adults who have not lost this quality but have used it to lead a satisfying life."

Although she found high school, in general, socially and politically isolating, Hunt was able to piece together a community of her own. She made friends with students in the book club, and outside of school her friends came from the Saturday Science Program and her church.

The Saturday Science Program at Columbia University that her teacher Charles Wilson had told her about provided the stimulation and scientific training that would influence the rest of her life. She began by taking two science courses, including one in astronomy. It was only when she ran out of science courses that she reluctantly agreed to take mathematics. "It didn't sound very interesting, but I did it anyway. That is definitely when

I changed my basic interest from chemistry to mathematics." She discovered that she loved abstract mathematics as she learned about groups and fields—a passion that she kept alive by reading mathematics books on her own throughout high school. These included the New Mathematical Library series that the Mathematical Association of America put out for high school and early college-level audiences—books such as *Continued Fractions, Numbers: Rational and Irrational,* and *Geometric Inequalities.*

During this period, Hunt read not only about mathematics but also about biology. She was excited by Mendel's genetics work on peas, and these ideas planted the seeds that later came to fruition in her research in mathematical biology. But at that point in her life, Hunt had no idea that she would end up as a mathematician; she simply could not envision herself in that role. "At that point I never really thought about what being a mathematician would be like—it was not something that I thought would be at all viable. Part of the reason was that somehow I got the idea that it was a rarefied kind of profession—that you had to be either wealthy or very, very smart, smarter than I was. . . . I really didn't see it as something that I could be doing on a day-to-day basis."

Although she could not yet imagine herself as a mathematician, Hunt *was* able to imagine being an electronics engineer. She persuaded her mother to buy electrical kits so that she could do some experiments at home, most of which ended up short-circuiting the house.

What, then, did finally enable Hunt to envision herself as a mathematician, to imagine that she could participate in this "rarefied kind of profession"? She remembers exactly what changed her mind. It was a booklet put out by the National Council of Teachers of Mathematics that had pictures and short biographies of several mathematicians, one of whom was Cathleen Morawetz, a mathematician at the Courant Institute, whom Hunt later met when she began graduate school at NYU. It was seeing a woman that made Hunt think that maybe it was possible to be a mathematician after all. Admittedly, she still had never seen or heard of any *black* mathematicians, but she was going to school during a period of great social change; in those days anything seemed possible.

College—Bryn Mawr

There was never a question in Hunt or her family's minds that she would go to college, but Hunt was not sure where to apply. So she sought advice from the Negro College Fund, which provided both career and college counseling. Hunt was given a list of possibilities, including many Ivy League schools, but she was most interested in the Seven Sisters colleges. She wanted to go to an all-women's college, but not one that was more like a finishing school than an academic institution. "I knew that I would have to go to the most academically demanding place because that is where I would get the best education." She chose Bryn Mawr because she was told that it was the most rigorous of the women's colleges.

Bryn Mawr was a total contrast to her high school. Where Bryn Mawr had fifteen hundred students (undergraduate *and* graduate students), Bronx Science had one thousand in her senior class alone. Bryn Mawr was all women, whereas Bronx Science not only was coed but was about two-thirds men. The graceful collegiate architecture and spacious campus of Bryn Mawr were strikingly different from the newly built modern architecture of her high school. She welcomed the new environment and was excited about starting college.

But her undergraduate years proved to be a mixed experience.

Hunt: There were things that were very positive about the college. There were a lot of very smart people; I learned a great deal from the other students. I learned about anthropology, psychology, things that normally I wouldn't have learned about at all had I gone to a large university. I probably wouldn't have known enough to have taken those courses, and I wouldn't have met the people who would take them. Going to a large university would have been like my high school experience; I would have continued in a certain groove. But as it was, in the setting of a small college, I encountered people who had very different interests and who were coming from a different social class and background. So that was very positive. I did a great deal of reading when I was in college—general reading. I wasn't that good a student.

MP: *They didn't reform you until graduate school.*

Hunt: That's right! I did work at my math courses, especially in the second half of my college career. I was very serious about that.

But about my other subjects, I have to admit I was not as good as I should have been.

Although the diversity of experiences was new and exciting, Hunt also experienced a sense of social isolation through much of her college years. It is hard for her to disentangle how much of that came from being black in a predominantly white institution and how much came from being less affluent than many of her classmates. But there was another big issue as well.

I think the biggest source of my unhappiness was the basic philosophy of the faculty at that time—their attitude was that if you had a question you wanted to investigate, you should find out what everybody else who had investigated that question said. Although I agreed, I thought it was important to start out fresh and see what *you* think first and then consult other people's work. I felt out of sync with the faculty.

Creativity

This independent streak and the tendency to run counter to the prevalent attitudes in her formal education are recurring themes in Hunt's life. She was influenced early on by a little book about creativity and science by A. D. Moore, whose thesis was that "creativity is one of the most important attributes that a scientist can have, and that sometimes education can really work at cross-purposes to developing a kind of freshness and independence of mind." As Hunt goes on to say, "I felt

that the faculty were promoting something that was quite the opposite. So I was rebelling against them."

How did this rebellion manifest itself in mathematics? "I tried to be a little bit unusual and inventive in terms of solutions to problems and things like that. I was always trying to look for the unexpected. I think most of the time I was simply wrong." The professors acknowledged, however, that while her approach was often not the most efficient one, it did display creativity.

Hunt learned quickly that a creative spirit necessarily entailed a willingness to be wrong. This has always been, and continues to be, an enormous challenge for her. "Willingness to look bad, to be wrong, is very difficult. . . . Things would have happened a lot faster had I been more confident about that. It took a very, very long time, and I'm definitely better than I was, but I still have reticence. You can't really make good things happen unless you make a certain number of mistakes. And you might as well go through them right away."

Hunt goes on to say that it can be particularly difficult for black people in the intellectual sphere because "if you make a mistake, then some people are likely to say, 'Well, that just goes to prove how stupid "they" are.' That's unendurable. Nobody wants to add to the stereotype."

What allowed Hunt to break through that barrier of fear of failure?

I think that my desire to become a mathematician—well, my desire to do research, something worthwhile in some sense—overcame (eventually) whatever reluctance I had. I never knew at any point before the end that I would indeed get a PhD. There never was a time when I thought, This is it! It's all downhill! Until the very, very end. But I knew that if I did not risk trying, it would be certain that I would not get a PhD or become a mathematician. So it was always balancing on the one hand the certainty of failure if I did not risk, against the uncertainty of success if I did risk. So I took the uncertainty.

Thinking about a Career in Mathematics

During her undergraduate years, Hunt read a book by Richard Bellman about the impact of computers on mathematics in biology. It had a formative influence on her later choices.

> The advent of computing made it possible to look at models of biological systems and analyze them. This was a new field, and he made clear it was a good field for people going into research because there were lots of problems, some unsolved for hundreds of years. There was room for people of all abilities to make contributions, so I took this to heart. I don't think I did anything about it immediately as an undergraduate, but it must have started me thinking. Eventually I did work on mathematical biology.

Throughout college, Hunt wanted to pursue a career in mathematics, but she had "grave doubts" about whether she was "really good enough to do that." So she wrote a letter to one of her professors in the mathematics department, Marguerite Lehr, an algebraic geometer and someone who Hunt felt was "an outstanding mathematician." In the letter Hunt wrote about her interest in mathematics and some ideas she had for research. She also wrote about her doubts in her own ability. The woman responded with a warm letter saying that Hunt *should* continue and that she was encouraged by the specific ideas for research that Hunt had mentioned. The professor said, "It's often the curiosity that one has about mathematics that is a sign that you have what it takes."

This was the encouragement and inspiration that Hunt needed to pursue a career in mathematics. She could have simply applied to graduate programs and assumed that if she got in, she was qualified to pursue mathematics. But what she needed was the personal encouragement of someone she admired and respected to bolster her confidence. This pattern of doubt and encouragement does not occur just once in a mathematician's life—it is a cycle that continues throughout one's career: during college, graduate school, thesis work, finding a job, getting tenure, and so on. Encouragement and support are needed at all of these stages for both men and women. But for women, and especially black women, these do not come easily.

Graduate School

Hunt's choice of graduate school was influenced very much by one of her professors at Bryn Mawr, Martin Avery Snyder, who had just graduated from New York University. He said it was a good place and that she

Figure 11.2 Hunt as a graduate student on vacation.

could get a scholarship to go there. Although she thought about other graduate schools, including Yale, she did not know anyone who had been there, and she worried about her chances of getting in. The Courant Institute of New York University, therefore, seemed like a natural choice.

Her experiences at Courant were again mixed; there were some wonderful aspects of the program, but her path was neither easy nor fluid. Although she received her master's degree, she also received a B instead of the required A on the qualifying exam and as a result lost her fellowship and her office. At the end of her second year, she left the university. She worked for a while until she was ready to return, at which point she received a fellowship from NYU, which enabled her to finish her course work and begin her dissertation. She received an office again, and she began working as a lecturer at NYU.

Graduate school was the first time Hunt really began to feel a sense of camaraderie with her fellow students. The faculty were focused primarily on their own research and did not spend a lot of time interacting with students. As a result, the graduate students banded together and formed a culture of their own. Hunt describes it as a "very jolly and supportive" atmosphere. Students regularly worked together, particularly in preparing for exams. They would take practice exams with each other, pretending to be the committee of examiners. Even after many of Hunt's peers had graduated, they would come back to be together. Although most of the graduate students were men, there was a critical mass of women that was sufficiently large to contribute to the positive atmosphere for women. A number of her peers became lifelong friends, mentors, and future collaborators.

One important source of support for Hunt during her graduate years was an organization of black graduate students called the New York Mathematics Society, later renamed the Baobab Society. It was formed to create a sense of community among budding black mathematicians who were spread out all over New York City. They met regularly for many years, holding seminars in which they would talk to each other about their research or bringing in other black mathematicians to talk to the group. An important function of the group was to support each other, "to be a sounding board for students, talk about mutual difficulties, and advise each other, especially those who were doing their theses. There were many friendships born then which have more or less sustained themselves."

A Support Network

Graduate school is a critical period in the development of a mathematician. This is where one forms a community that will be crucial throughout one's professional life, and this community plays a powerful role in shaping one's sense of self as a mathematician. Having this kind of community is an important source of support and camaraderie and yet is often so taken for granted that it becomes invisible. But for women, particularly women of color, such a sense of camaraderie and community is not as easily formed. When Hunt found it both among her fellow graduate students and in the New York Mathematics Society for black mathematicians, she had a support network that could help her through the obstacles and difficulties one inevitably faces as a graduate student. And although there were excellent teachers at NYU, it was really the student community that she learned the most from and that defined her experience there.

Having such a community made it much easier for Hunt to imagine herself as a mathematician. She was no longer alone: as a woman, as an African American, as a young struggling mathematician. The connections she formed through these graduate student communities helped her to find a thesis advisor and a job when she graduated and helped to establish her research affiliation with the National Institutes of Health.

One of Hunt's fellow students heard her talk about her interest in mathematical biology, and it was he who put her in touch with the man who became her advisor, Frank Hoppensteadt. Hunt describes her advisor as an unusual person because he is a mathematician and at the same time is very interested in science. Both his style and his interests dovetailed well with Hunt's needs.

He had confidence in me. And he was a strong and good enough person to say so. He didn't say it often, but he said it. It's what I probably needed to hear. I came to him with a lot of skills. I think I was twenty-seven years old when I started working with him, so I was a little older with a lot of math under my belt, so I wasn't green in that sense. Even so, he somehow seemed to be able to pace me. He knew how much to give me and how much I could do, but he did not baby me. He didn't give me any breaks when it comes right down to it. He gave me a chance, and he expected me to perform, and he said I was capable of doing it and capable of pursuing a career in mathematics. He treated me like a professional, a fellow professional.

She had found an advisor, but the process of writing a dissertation was quite challenging. She was working on difficult problems, and there were very few people working in the field of mathematical biology to turn to for help. Fortunately her advisor had a good sense of how to keep her going. If she got stuck on a problem for too long, for example, he would redirect her toward a new problem, so that she would stay mathematically active. But in her last year of graduate school, a different kind of obstacle arose. Her advisor left NYU and went to the University of Utah—not a place many black people would have chosen to attend. Nonetheless, she did follow him out there for a year to complete her thesis, taking a position in the mathematics department. To her surprise, it was a comfortable and pleasant place to be. Piecing together her solutions to problems her advisor had given her, she finished her thesis midway through the year, and came back to NYU to defend it over Christmas—a major hurdle conquered.

Howard University

While Hunt was at NYU a man named Jim Donaldson came as a postdoctoral visitor. He was the chair of the mathematics department at Howard University, and he invited her to come visit Howard, which she did. It was a good visit, and he said she should look them up when she was ready for a job. That is exactly what she did, and they hired her immediately.

Hunt did not learn until much later that other universities were interested in her as well. Later, these schools said they did not understand why she had not applied, but she was given no indication that there was any reason to apply to them. Hunt believes that in order for affirmative action to work, schools must take a more active role in seeking out women and minorities and encouraging them to apply for jobs, just as Jim Donaldson did to lure her to Howard. At the time, there was nothing obvious or natural about applying to a host of unknown schools.

As it turned out, being at Howard was good for Hunt in many ways, especially because of the tremendous support from Donaldson.

After I finished my dissertation, I had some surgery, which should have been fairly routine, but the upshot of it was that I was allergic to the anesthetic, and I got toxic hepatitis. The stress of finishing the dissertation and the illness was like fighting and squeezing my way out of a very narrow, tight opening. It left me tired and a little uncertain about the future and whether I would be able to do any research. I did start working again, but it was Jim who always had a high opinion of me, and I appreciated that. He would say, "Don't worry; you'll find a problem and you'll do it!" He just took that for granted.

And indeed, Hunt has managed to continue to be productive in her research even with a fairly heavy teaching load at Howard. She had leaves to pursue research at the National Institutes of Health and later at the

Figure 11.3 Hunt with a student after teaching a class at Howard University.

National Institute of Standards and Technology. She attributes her productivity both to the fact that she has very few family demands and to the fact that she has had so much encouragement in her research from people such as Jim Donaldson, Adeniran Adeboye, Tepper Gill, Gerald Chachere, and many others on the faculty at Howard University.

The Role of Religion

Like every mathematician, Hunt has her good periods and her discouraging periods. Mathematics can be hard, time-consuming, and frustrating. "Sometimes I go through long periods where nothing really works out." What keeps her going during these times is her love of mathematics and her religious faith.

Many scientists who have a spiritual dimension to their lives are reluctant to talk about it publicly because modern-day science is usually pitted against religion as if the two were contradictory. This is a relatively modern view, and many individual scientists find the blending of science and religion natural and important.

Although Hunt was raised as a Christian and actively participated in church, during her late high school years she found herself becoming more disillusioned, and she finally became an atheist. In college she met what she describes as "a couple of the very few people on campus who not only went to church but really believed."

They didn't proselytize, and in fact you had to know them well before you figured out they did go to church. They were full of

jokes and puns—they were quite irreverent and at times mischievous. They gracefully blended these qualities with courage and integrity. I thought this was so odd; I knew very few people like this. Since I found the whole subject of religion to be embarrassing and beneath notice, I tried at first to overlook this weakness in them. But they were good friends to me and I really liked them, so after a while I started thinking.

Then the summer of her sophomore year, Hunt had an experience that fundamentally changed her life. With the help of the dean of the college, she was offered a summer job as an engineering assistant in New London, Connecticut. The dean tried to set up housing for Hunt, but it fell through. So when she arrived in New London, she tried to find housing on her own. But no one would rent her a place to live because of her skin color. It was the first time she had encountered such overt and direct discrimination. She was eighteen years old.

Hunt: It was not subtle and it sort of sent me reeling. I got pretty close to the edge then. I was by myself. My folks were in New York. There wasn't anybody I could go to. Somehow I needed help to survive this. It was shattering. For the first time, I understood that religious faith can sustain you. It is the real ground on which we stand—not our family and not our friends, however much they love us. I wish there were less painful ways of finding this out. As it was, I am grateful that I was "brought up short" on this issue early in life. After many years, I can say that the most important gift that religious faith has given me is gratitude for being created the way I was. Perhaps I wouldn't have been happier, and I almost certainly wouldn't have been a better person, had I been born with qualities that some view as more "acceptable."

MP: *Did you know that discrimination existed when you were growing up in New York City?*

Hunt: Oh, yes, I knew that discrimination was around, and I saw it, but I never understood the extent of it. At home, I could always go back and tell my family about it. I had this matrix of support. During high school you saw discrimination, but also, at that time in New York City, there were still lots of people of good will who were committed to integration of some kind, and the social fabric had not yet pulled apart, and the economic conditions had not worsened to the point where there was really overt racism. In some parts of New York City, things were bad, but you just avoided those places. In a completely new environment, though, I had no guideposts—college, after all, had been rather tolerant. Now there was nobody I could go to, and I couldn't find a place to live.

MP: *Total isolation.*

Hunt: Yes, totally isolated. I was living in a place where no one would speak to me. I was it. My choices were to turn around and go back on the train or to survive. And I couldn't go home. Maybe that was sort of foolish, but I felt I couldn't go back to New York. I needed the money and this job.

MP: *So you stayed?*

Hunt: Yes. And I eventually got acceptable housing. I think I called the NAACP, and they referred me to somebody. I think that's eventually how it worked itself out. But even at work the people weren't terribly forward. I have a fairly thick skin; that, plus being at home, always provided this cushion, but suddenly it was just stripped.

MP: *When you later experienced discrimination, was it less difficult because you had already been through that?*

Hunt: I learned to take a long view. A couple of decades down the road, these things are going to seem trivial. You have to think about the end of your life. If you sum up your life, what do you want to say? When the curtain comes down, what do you want the review to be? So several things helped me: I took a longer view, I had inspirational resources, and I also have a very stubborn personality that says, Well, I'll show them!

For Hunt, dealing with doubt and uncertainty was an integral part of her life. It brought together questions of who she was as a person and as a mathematician.

The crisis of youth is the crisis of self-definition. Even in a society where an individual's role is defined early and very clearly, there is still the uncertainty of not knowing at a fundamental, spiritual level why we are here and what our ultimate purpose in life is. In a society as fluid as ours—where small differences in our choices can have large consequences—the question of definition is that much sharper. Problems of race and gender discrimination posed challenges to who I was at a fundamental level. I was forced to search for answers at that level. I think this happens to anyone who has experienced suffering over a long period of time.

Given how important religious faith is in Hunt's life, how does she blend her mathematics and religion? How does she reconcile what some see as contradictory enterprises? For Hunt, the belief in realities beyond our concrete physical world is what unites the mathematical and spiritual components of her life.

As a mathematician you are dealing with abstract structures, and the kinds of things that impressed me most were not really visible. So as a mathematician it is easier to believe that the kinds of concrete things that we deal with every day do not constitute all of reality; there are realities that are not immediately perceptible by us. That's the basic difference between someone who is some kind of a theist and somebody who is not: the belief that reality is just far wider than simple physical laws, economic and political laws, or relationships that we see.

Lessons Learned from Teaching

Hunt was motivated to teach by a desire to convey to the next generation life lessons and inspiration. "I have learned some things. I've read a lot. I have insights. I would like for the next generation to know about them. As a human being, there really isn't any better purpose for anything you do than to try to give other people the benefit of what you've learned, so that it's a little bit easier for them to advance. Teaching gives me an opportunity to do that."

In giving talks at a variety of schools around the country, Hunt always finds ways to describe how interesting mathematics can be. Often the students ask questions like "What is it like to be a scientist?" Hunt responds: "I tell them that it brings me a great deal of personal satisfaction. Many people feel a lack of direction that comes from not having a strong vocation or a sense of calling to do something. Income, position, being able to support a family at a high level can be motivation for a career." But for Hunt, there is a very deep level of satisfaction in a life of the mind. For this reason she tries to convey that mathematics is about "how to deal with ideas and put together ideas that have some structure and to have an appreciation for them. The world is bigger than what job you have, your income, what car you're driving, or where you live—that's something you try to hold out to people of all ages, but especially younger people."

Three Life Lessons

There are three life lessons in particular that she hopes to convey. The first is about gaining confidence. "I would like to convey to students a little more confidence in themselves

Figure 11.4 Hunt working with students.

order for a teacher to convey that to students, the teacher has to believe in them. She cites the Helen Keller story to illustrate that "in order to be able to teach a person, you have to take them seriously. You can't have contempt for them."

In general, she says, "I hope to convey a certain amount of independence of mind. That is the single most important thing, that when I'm not there, they will be able to approach a problem and be able to think about it. It's a habit of mind, to be able to think about a problem, come up with a hypothesis, and be able to move toward a solution." She tries to communicate how mathematics is not in some distant world, but rather "pops up right in front of you." But the problems are disguised, so you have to recognize them in order to work on them. Otherwise "you're helpless in that situation."

and their ability to be able to deal with difficult problems, to know that, at least to some extent, some sense of self-mastery is to be gained by studying."

Second, she hopes to communicate the pleasure of intellectual activity. It's really quite a great life that way. It involves "coming up with the unexpected, enjoying challenges, and having a feeling and appreciation for structure." She goes on to say, "It shouldn't just be left to artists. It's something that other people should have, in particular mathematicians."

The third lesson she tries to impart is that blacks can and should do mathematics. But in

Encouraging More Women and Minorities in Mathematics

Hunt believes strongly in making the field more inclusive—that one's sex, race, or class should not prevent one from studying or pursuing mathematics. She uses an analogy to basketball to emphasize that there are many levels of talent.

Math is a lot like sports. There is a lot of talent and many kinds of talent. Take basketball as an analogy. We all appreciate players like Larry Bird, Bill Russell, Bill Walton, as well as Nate (Tiny) Archibald or Isiah Thomas. They have (had) very different abilities and styles of play. Yet they were all marvelous players. So the pool of talent is broad. And when you consider Patrick Ewing, Magic Johnson, Julius (Dr. J) Erving, and finally Michael Jordan, you also see an almost infinite depth of talent. There is probably a broad consensus that Jordan is the greatest player the game has produced, but does anyone think that diminishes the contributions of Julius Erving? Would Jordan have achieved as much without the help of the less talented Scottie Pippen? Mathematics is like that. No matter how good you actually are, there is definitely somebody who can run rings around you. If you encounter these people, it can be intimidating. This, with the fact that mathematics is a field that a lot of people have trouble with, causes a great deal of anxiety both within and outside the profession. I think we should minimize it by trying to be a little more inclusive, by trying to look for the talent people *have,* rather than dismissing them for the talent they lack.

Encouragement is crucial in developing mathematical talent. Hunt gives an example of a student in one of her courses who was doing nothing; the highest grade he got on an exam was a 70. But she noticed that he was going in regularly to see one of her colleagues. Her colleague said, "Yeah, he's a lazy guy, but he is bright" and therefore was giving him some hard problems to work on on his own. The student was in fact doing very well on them, so Hunt decided to follow suit and had the student do an independent project. Not only did this engage him in that work, it improved his participation and performance in class as well.

Hunt: It's important especially when you're trying to increase the pool of so-called underachievers [to find ways to encourage them]. There are a lot of reasons for poor performance. There are more reasons for poor performance than there are for good performance. It makes things much, much more complex.

MP: *Like they might just be unwilling to speak up in class or . . .*

Hunt: Yes, or they feel that they can't speak up in class. It may be that they don't have good study habits or that they don't know how to work with other students or that other students refuse to work with them. There could be other reasons, subtle and difficult ones having to do with personal issues like lack of maturity or family conflicts.

Even in her own undergraduate institution, Hunt was disturbed by the attitude of the mathematics department, which focused only on those few students who were going on to become research mathematicians and ignored the rest. She sees the same phenomenon in graduate programs that turn away talented graduates of small liberal-arts schools because their background is not as comprehensive as that of students coming from large, prestigious universities.

Figure 11.5 Hunt receiving the Arthur S. Fleming Award for Outstanding Achievement in Science in 2000.

This is especially disturbing because some of the small liberal-arts schools have produced a disproportionately large number of female scientists. If top graduate programs close their doors to these students, they are turning away a significant pool of potential women and minority scientists.

Ultimately, Hunt believes that the most important quality for success in mathematics is intellectual curiosity. Clearly a small school cannot offer the range of courses a university offers, but Hunt believes that "if a major can graduate with a solid understanding of advanced calculus and abstract algebra, I think you've got a very good product there, and I think that is somebody you can definitely work with." She argues that various mechanisms or programs could be developed to ease the transition of students from four-year colleges to graduate programs.

In summary, Hunt says, "Our methods don't pick up every possible aspect of the kinds of talents that could potentially be useful. And it's for that reason that we should be more inclusive. Our tools for discerning talent are blunt and imperfect."

Recommendations

Hunt has many ideas for ways to make mathematics more enticing for women and minorities. The following is a list of general ideas that can be incorporated into a classroom, followed by a list of more specific techniques for encouraging women and minorities.

Present Mathematics as a Human Endeavor

Hunt argues that "mathematics history is not taught at all, and I think that's a mistake—math history was something that I had read in high school, and without a doubt it helped me make the decision to become a mathematician." She talks about how inspiring it was to read about the lives of mathematicians: "It makes the kinds of work they were doing exciting—something really worth aspiring to. By being presented in the context of its history, the subject matter becomes more human, thereby making it more appealing to a wider range of students."

A second way to reveal the human dimension of mathematics is to talk about the people behind the mathematics. Hunt makes a point of talking about David Blackwell, who developed some of the material they covered in her classes, when he also was chairman of the mathematics department at Howard for ten years. Since he is black, she thinks it is especially important for the students to hear more about him than his name.

A third way to present mathematics as a human activity is to try to convey the interconnections between a society and mathematics. This is a topic that Hunt finds fascinating and continues to learn about. When we spoke, for example, she had recently been reading about the French Revolution and how important it was in the development of mathematics and technology in France.

Use Real-Life Examples

Because Hunt often teaches probability and statistics, she tries to find examples that could emerge from everyday life. Medicine abounds with basic statistics and probability problems. She might present, for example, a hypothetical scenario. Suppose a doctor decides to give a transfusion to a patient who is a hemophiliac. There are two pools of blood in this country, one obtained from volunteers, and the other obtained from those who donate blood for money. The latter is much more likely to have diseases such as hepatitis. The doctor might decide to use the second blood supply, which is normally used only in extreme emergencies, arguing that the probability that the patient has already been exposed to hepatitis is 100 percent for the following reason: If the patient is a hemophiliac and has already had 100 transfusions, and the probability that a transfusion contains hepatitis is 1 percent, he might multiply 100×0.01 and get a probability of 1, that is, a 100 percent likelihood that the patient has already been exposed to hepatitis. But this would be a drastic miscalculation. The real probability is $1 - [0.99$ to the 100th power$] = 63$ percent. This kind of mathematical understanding is critical to medicine as well as many other types of professions, including law and business. Whenever possible, therefore, Hunt talks about her own work, why it is important, and how it relates to the real world.

Ultimately Hunt sees mathematics as a deeply human activity that gives us a broader vision of ourselves and helps us find meaning in our lives. At the same time, she sees it as a luxury that must be appreciated and passed along, or it could be lost. As she so eloquently states,

> The point is that as soon as there was enough to eat and the environment was

relatively stable, humans were involved in mathematical activity. But this kind of broad vision of mathematics is important in order to secure its future. There is nothing that says that mathematics ought to continue as a cultural activity. Indeed, there are many arts and crafts that have been lost as the civilizations that invented them declined. The first six books of Euclid's *Elements* were lost in the fire at the Alexandria library—a significant blow to the development of mathematics. It would be terrible to think of something like that happening now, but it could. It is better to invest in as many people as possible—to convey the idea that science and mathematics elevate the mind and the spirit and that they are something that people will always need. Somehow that's not being transmitted. Somehow we're letting our machines, our greed, and our own spiritual emptiness devour everything. So the enterprises of teaching and research in mathematics—whether in academia or out—spreading the knowledge we gain to as many as possible are not only noble activities, they are also the conservative thing to do.

The specific techniques Hunt uses to create an atmosphere of inclusivity in the classroom include the following:

Extensive Use of Questions

Most of Hunt's classes involve a kind of call-and-response pattern. She makes a point of always starting with simple questions to get the students initially engaged. Once they feel comfortable thinking and talking out loud, they are more willing to try ideas they are less sure of and are more willing to be wrong in class. If the answer is even remotely right, she tries to work with it.

Since the males in her classes tend to speak more and have more confidence, she makes an extra effort to involve women and hear what they have to say. She finds that this little bit of extra attention is quite effective in drawing women out.

Encouragement

Hunt believes that the most important thing of all is to encourage students. Everything else pales in comparison. This is especially important with any student who expresses interest in mathematics; and since some may express it in very quiet ways, it involves paying careful attention. For example, a student might attend math club meetings but never speak up, or may be very attentive in class but shy about answering questions. So how does one encourage such students? According to Hunt, "I think the most important thing would be to gradually, without pouncing, try to gain their confidence."

> People love to be flattered. Don't become unrealistic. But flatter them. Compliment them. Because I can assure you that if you're a white male faculty member, you probably are not really aware of the extent to which black students are not complimented. Complimenting encourages. Don't be extravagant to the point that it's ridiculous, but be encouraging. If you can, gain that student's confidence. Try to take a genuine interest; people will see that, and they will open up a little

bit. This may seem intrusive, but will be appreciated.

Role Models

It was not unusual for Hunt's students to tell her how much they appreciated having a female black instructor. At the time of the interview, she was the only African American woman in the department—despite the fact that Howard University is a predominantly black institution! Hunt was written up in a brochure that was used in an exhibit in a Chicago museum. From this she has gotten

Figure 11.6 Hunt in her home office.

numerous letters from students who were interested in mathematics and had never seen or heard of a black woman in mathematics; they were hungry for a role model.

Some people object to the idea of a role model because they think that each person is unique, and no one can be a role model for anyone else. But a role model, more than anything else, simply opens doors of possibility. As we see in Fern Hunt's case, for women or minority students who have never seen a mathematician of their gender, race, or ethnic background, it can be crucial to find such a model of possibility.

Education in Black and White

Race has obviously been an important factor in Hunt's education and development. How has this influenced her thoughts about the advantages and disadvantages of predominantly black schools versus predominantly white institutions?

MP: *If you had children, would you advise them to go to a predominantly black college?*

Hunt: Not necessarily. A lot would depend on the personality of the student. There was a point, when I was teaching at Howard, when I would say that they should definitely try to get into Dartmouth College, for example. At that time there wasn't the racial exclusion that there had been in previous years. The majority of colleges were integrated. They were admitting students of all colors and actively seeking black students. I advised students

at that time to try to go to majority colleges, especially with their previous experience of being in predominantly black schools. You need to meet other kinds of people; I felt that was very good and very healthy. We need to do that as a society, we need to get out of our collective ghettos. However, in recent years there has been a growing intolerance, and I have met some of the students who have transferred from majority colleges to Howard. There is a terrific economic anxiety among white students right now, and some of them seem to be taking it out on black students on these campuses.

I also find that sometimes black students are forming little cliques on the majority campuses. So the situation is very complex. I no longer hold the unequivocal position that yes, you should definitely go to the majority, or yes, you should definitely go to a black college; there are pluses and minuses with both choices. It would depend very much on the individual.

In the end, Hunt says, there is no clear answer to this difficult question. It depends entirely on the particular student, his or her interest and personality, and the particular institutions in question.

Clearly Fern Hunt is just one person, with one set of experiences. But her story illustrates some of the ways that being a woman and being black influence one's mathematical development. Being black affected Hunt's experience at Bryn Mawr, which, while all female, was a predominantly white institution; it influenced the positions that opened to her when she completed her PhD; and it influenced her experiences with her students at Howard, a predominantly black institution.

At the same time, certain lessons emerge that are universal—most notably that support and community are critical factors in achieving success. This is a lesson that has influenced Hunt's style in teaching. It also suggests one direction for improvement in making mathematics a more inclusive endeavor.

Postscript: On page 211, in the reference to the "loss" of the first six books of Euclid, many details in the complicated tale are left out. Readers interested in a fuller account of the history of versions of Euclid available at various times in history may wish to consult Florion Cajori's *History of Mathematics, Fifth Revised Edition*.

Twelve

Dusa McDuff

Donald J. Albers

Dusa McDuff is a highly accomplished mathematician who works in symplectic geometry, a relatively recent and somewhat esoteric branch of mathematics. She says, "Symplectic geometry is an even-dimensional geometry. It lives on even-dimensional spaces and measures the sizes of two-dimensional objects rather than the one-dimensional lengths and angles that are familiar from Euclidean and Riemannian geometry."

McDuff claims that if it had not been for Miss Cobban, her high school math teacher who taught her geometry and calculus, she might not have become a mathematician. She did know that she had to do something to impress her father, a geneticist and developmental biologist, and fulfill the ambition of her mother. She also felt that she needed to live up to her grandmother, Amber, who as "Dusa" had a scandalous affair with the famous author H. G. Wells.

Her mathematical success is shown both by her election in 1994 as a Fellow of the Royal Society of London, the second female mathematician to be so honored after Dame Mary Cartwright's election almost fifty years earlier, and also by her later elections to memberships in the National Academy of Sciences (U.S.) and the American Academy of Arts and Sciences. She now holds an endowed professorship at Barnard College, a college of Columbia University.

Her PhD thesis in functional analysis from Cambridge University was published in the *Annals of Mathematics*, one of the most prestigious mathematics journals in the world, a rare event for any doctoral thesis. A short time later she went to Moscow where she came under the influence of the famous mathematician I. M. Gelfand. Over the next six months she learned much from him and returned to Cambridge determined to work in a new field. In the interview that follows, McDuff describes the difficulties in shifting fields, including the special problems that a woman encounters in succeeding as a mathematician.

Figure 12.1 Dusa about 1948.

MP: *Thanks very much, Dusa, for meeting in your MSRI [Mathematical Sciences Research Institute] office today to chat about yourself and how you became a mathematician. Let's start at the beginning. I understand you were born in October of 1945 in London, and moved to Scotland when you were a child.*

McDuff: Yes. My father, Conrad Hal (Wad) Waddington, got a professorship at the University of Edinburgh in 1947 when I was two. A geneticist and developmental biologist, he worked during World War II in Patrick Blackett's lab devising strategies to defend against U-boat attacks. He told me that this group started the field of operations research but couldn't publish until the 1970s because of secrecy laws. When the war ended he

traveled extensively on war-related work. Demobilized in 1947, he took up his position at Edinburgh as Director of the Institute of Animal Genetics.

Mortonhall—An Amazing House

After the war, there was a serious housing shortage in Britain. My father's solution was for everyone connected with his lab to live together. We moved into Mortonhall, a large stone mansion on the outskirts of Edinburgh, whose grounds had statues, clipped yew hedges, a large walled kitchen garden, and stables. The stables had no horses, but I remember collecting the chicken eggs. Although people lived in separate flats and rooms, we ate communally, with one dining room for the grownups and one for the children. Rations were shared. Much of the work of the house was done by a household staff that (at least in the later years when I got to know them) included many displaced people from countries like Estonia and Latvia. My father set up the household at Mortonhall almost as a social experiment; my mother, trained as a town planner, was always very interested in the influence of architecture on communities.

Although these arrangements suited our family very well since my mother worked full time as an architect in the Scottish Civil Service, they made some of the other wives rather unhappy. There were many other tensions in Mortonhall. It was different for the kids, but for the grownups, it was not necessarily so easy to live that way, partly because there was no escape from the lab but partly also because life in Britain at that time was very austere. There was little food, and houses were cold because of lack of fuel. Children got rationed orange juice and milk and other things like cod liver oil, but grownups didn't.

Squabbles over Porridge

Edith Simon, one of the wives, wrote a novel about life in Mortonhall called *The House of Strangers*. One of the few things in it that rang absolutely true was stories of the grownups at the breakfast table squabbling over the porridge. There was a porridge rotation and scientists are not necessarily good cooks. They took the precious oatmeal and made porridge that was lumpy, or weak, or burnt, or too salty. Then they had to share their meager butter and sugar rations and would look to see exactly how much their neighbor took. It was difficult.

MP: *It sounds that way.*

McDuff: But it was wonderful for the kids.

MP: *How many kids were in the house?*

McDuff: Well, counting myself and my sister Carrie—I guess six or eight. My mother worked from nine to six every day and a half-day on Saturday, and couldn't come home at lunchtime because her job was so far away. My sister and I went to nursery school and had Irish nannies. But for a lot of the time we were running around in gangs outside with very little supervision, able to

Figure 12.2 Dusa and Caroline playing in Mortonhall, about 1948.

Figure 12.3 Dusa (on left) and her sister Caroline, about 1949.

do what we wanted—climbing trees, playing hide and seek. I also remember talking to the cook in her enormous old fashioned kitchen, the boiler man as he stoked the basement furnace, and the gardener working in the kitchen garden.

MP: *It certainly sounds like a memorable childhood experience.*

McDuff: Yes.

MP: *How long did that last?*

McDuff: Five years.

MP: *That's a big chunk of one's life at that stage. You moved there when you were two.*

McDuff: Yes. At that time it was unusual for young children to go to nursery school.

Early Interest in Math and Grandfather's Influence

My mother, who cared deeply about our education, chose a very progressive school run by the Parent's National Educational Union (PNEU). I used to love school, especially doing math. I was allowed to do whatever math I wanted. By the time I was seven, I was way ahead of the others.

MP: *In one of your writings, you mention the influence of your grandfather on your mathematical development. You said that he had done math before turning to the law.*

McDuff: My grandfather G. R. (Rivers) Blanco White was the Second Wrangler at Cambridge (i.e., he placed second in his class) in 1904 and eventually became a divorce court judge. A private in the First World War, he served in the artillery, using his mathematical knowledge to calculate trajectories.

MP: *So, he influenced you when you were quite young.*

McDuff: I met him once when I was four and then again when I was about eleven. We were not a close family. So, this visit when I was about four was very special. I remember him showing me the multiplication tables.

MP: *Were your mathematical interests apparent before he told you about multiplication tables?*

McDuff: Probably. Since I liked math, I certainly knew how to add and multiply. But I had never seen a full ten by ten multiplication table before he showed it to me, explaining its various symmetries and patterns, how if you look down the nines column, the numbers change regularly, one digit going up each time and the other going down. He showed me its beauty.

MP: *You said that a nice aspect of the school was that you were able to do as much math as you wanted. Do you recall what it was about math that was so attractive to you when you were little?*

McDuff: I just loved mathematics, I don't really know why. My mother was good at math (Rivers was her father), and she was always eager to encourage us intellectually. She told me once that she was thrilled when my "first word" was two words with two ideas. I was a bit precocious and quick at doing some things. I liked the way my sums came out correct. I remember doing an entrance exam when I was six to get into a proper school. They asked me to add two and three, or something like that. I said, "This is far too easy." I did a sum with four digits, and all the teachers gathered round astonished.

Figure 12.4 Dusa with "Mousie" (age ten).

Figure 12.5 Dusa aged about sixteen.

MP: *So, your interest in mathematics was early and strong, and it's never abated.*

McDuff: I've had other interests. I always wanted to be a mathematician (apart from a time when I was eleven and wanted to be a farmer's wife) and assumed that I would have a career. Luckily I had a very good math teacher in my high school. I went to the same girls' school from seven to sixteen, the best my parents could find in Edinburgh. I despised the science teachers because they could not answer my questions, but I respected the math teacher Miss Cobban; she taught me Euclid and calculus. Otherwise, I don't know whether I would have become a mathematician.

MP: *What careers did your siblings pursue?*

McDuff: My sister Caroline Humphrey is Professor of Social Anthropology at Cambridge University and a fellow of King's College. My half-brother, Jake Waddington, whom I hardly knew while I was growing up, is an astrophysicist.

Father Thought Mathematics Was Boring

MP: *You said that your family very much valued creativity, and yet you always felt that in their view the really creative people were males.*

McDuff: My parents would never have said that explicitly. Unusually for the time, I was brought up to think I would have a career and that women could do just what men do. But my mother had also subordinated her career interests to those of my father, justifying that by the fact of his brilliance and the needs of her family. When I became a teenager, about fifteen or so, I felt that the kind of intelligence I had did not count for much, and what was really creative was a more artistic kind of talent. I might be very good at reasoning, but that was ultimately not important.

I had some artistic interests; I played the cello, and I loved reading. My boyfriend at the time, David McDuff, who became my first husband, was a poet and linguist. (He knows an astonishing variety of languages, including Russian, Finnish, Icelandic, and the computer language C++.) We met through music. I thought that he had a brilliant, creative mind, while I didn't see myself as creative. For example, although I was quite good at painting as a girl and now get great pleasure from going to art galleries, my sister was much better at painting than I was, with a wonderful sense of design and color. Although some people suggested that I study to be a cellist, I decided not to because I felt I had more talent as a mathematician.

MP: *So, although you had these strong mathematical interests and were doing very well at it, there was no apparent feeling by your mother and father that mathematics was a particularly creative area?*

McDuff: My father didn't like mathematics; he thought it was very dry. When I was thirteen, we had many conversations about the book he was writing called *The Ethical Animal*, about the development of the moral sense in humans through evolutionary processes. He gave me philosophy to read, along with *The Voyage of the Beagle* and Freud, greatly broadening my outlook. He prided himself on being a scientist, a philosopher, and an artist. He wrote a book about modern painting, *Behind Appearance*, that even today some people find worthwhile. He knew Alfred North Whitehead and Bertrand Russell from Cambridge in the 1930s, but knew Russell as a philosopher, not as a mathematician. His attitude about mathematics

Figure 12.6 McDuff's father, Wad (Conrad Hal Waddington) (about 1965).

was that it was boring—though in his later years he was very interested in developing a theoretical approach to biology and was open to the importance of mathematics in that connection.

My mother was an architect. For her, the artistic side of it came through design. She was passionate about research and pure thought (mathematics did qualify there!), but she didn't know enough about mathematics to emphasize its creativity. I first got to know people who I thought were truly creative mathematicians when I went to Moscow in my third year as a graduate student. I discovered there that mathematics could grow and develop. Before I had seen its compelling beauty, but it was somehow static; I was unaware of how it was created.

"I Had to Do Something to Impress My Father"

MP: *When you were at Cambridge, you wrote a thesis that ended up being published in the* Annals of Mathematics. *That isn't chopped liver, as some would say.*

McDuff: Right.

MP: *It got a fair amount of attention in the mathematical community.*

McDuff: I wasn't so aware of that because I changed fields so quickly.

MP: *You must have felt very good about it at that time.*

McDuff: Yes, but I was totally divided. I was deeply in love with David, a poet, and he was math-phobic. I had a separate life as a mathematician, with no mathematical friends. When I was an undergraduate at Edinburgh, I didn't talk to any of the other students. I didn't know any of them, except I remember playing bridge with them one afternoon.

My only friends, and I had very few, were through David, and they talked about poetry, art, and politics. I was learning Russian and German to keep up. When I was little, everybody thought I was brilliant; I got a lot of attention for always coming out top on exams and that kind of thing. At some point, I decided that was irrelevant, and I turned

my back on it, trying to live a different life. For one thing, I had to do something that would impress my father, and mathematics was not it.

"I Had to Live Up to My Grandmother Dusa"

McDuff: For another thing, I had to live up to my grandmother Dusa. I don't know if you are aware of that aspect of the story.

MP: *Well, you did mention in one of the articles that you have written that she was apparently very colorful and that she had a long, somewhat sensational affair with H. G. Wells. She bore a daughter, Anna-Jane, by him. Wells was a well-known writer then, and certainly his fame persists. Apparently Dusa was the name that Wells gave her.*

McDuff: The story that my mother told me was that he gave her that nickname because of her long black snaky hair. Then, later, I read about her in H. G. Wells' own book, *H.G. Wells in Love: Postscript to an Experiment in Autobiography*, a book about all the women with whom he had had serious affairs. One of the chapters is about my grandmother, Amber Pember Reeves. (She was the model for his very appealing "new woman" heroine Ann Veronica.) Her father, Pember Reeves, was the Governor General of New Zealand and then the first director of the London School of Economics. Wells said that Dusa was her private name for herself, chosen because she was fascinated by the image of the Medusa head held by Perseus in Bernini's bronze statue. I much prefer that version of the story.

Wherever the name came from, it is somewhat puzzling. As far as I knew, the name Dusa meant a terrifying monster that rendered others powerless. Recently, I discovered that in Turkey she is a guardian figure, her head often portrayed on Athena's shield; in prehistory she must have been an earth goddess because Medusa is the feminine form of the name Medon which means ruler. My mother never told me about those aspects of the name.

MP: *It's certainly a distinctive name.*

McDuff: I felt I had to live up to it. Being called after Medusa made me feel unique. The other schoolgirls made fun of me, pretending that, like the mythic character, I would turn people to stone if they looked at me. And then, as I realized much later, I thought I had to do the equivalent of running off with H. G. Wells: I would not be able to hold my head up if I didn't.

MP: *I have read a bit about your grandmother in* Shadow Lovers *by Andrea Lynn. She portrays her as a progressive feminist and important author of books on social issues, whose admiration for Wells continued for more than thirty years after their affair commenced in 1907. She quotes from a letter that she wrote to him in 1939: "What you gave to me all those years ago—a love that seemed perfect to me, the influence of your mind, and Anna-Jane—have stood by me ever since. I have never for a moment felt that they were not worth the price." I'm beginning*

Figure 12.7 Amber Pember Reeves.

to understand why you hold her in such high regard.

McDuff: At that time, the early 1900s, women were only beginning to have careers, and it was very hard to have both a family and a career. You had to choose. She had a job in the First World War in the Ministry of Labour promoting women's employment but had to leave it at the end of the war. She then wrote some novels and a book on economics for the Left Book Club as well as helping her husband "nurse" a constituency for the Labour party (i.e., stand for Parliament in a hopeless seat). After more war work during the Second World War, she became interim president of Morley College (part of London University for adult education) for a short while until an unmarried woman could take it over. She served there as Tutorial Lecturer in Moral Science for many years, giving evening courses in psychology and writing a book *Ethics for Unbelievers* that used Freudian principles to argue for a kind of Confucian restraint. Despite all this accomplishment (much of which I learnt about later), my family's attitude was that she had not lived up to her potential.

MP: Did you ever meet her?

McDuff: Oh yes. I met her several times, not very often when I was small, but when I was a teenager, I spent a few days in her house. We talked quite a bit then, and we also corresponded for a long while.

MP: What do you remember of your personal interactions with her?

McDuff: My grandmother liked telling slightly risqué stories about the people she knew. She was very good at that! She spoke in incredibly complicated well-formed sentences of the kind nobody uses now. She knew Lloyd George, Beatrice and Sydney Webb, and many other politicians and socialists. She once told me a story about Beatrice Webb getting a bee in her blouse at some garden party, which made her lose her poise. Beatrice was known to be so very correct.

We also talked about what I was doing. She liked me. I remember one day as I left the room I heard her mutter "she's my favorite granddaughter." She wouldn't tell me that to my face. I really enjoyed my relationship with her. I wrote to her, and she regularly wrote back. As she got older, my mother wrote a letter to her every week.

MP: Wow. A letter a week!

McDuff: People did write letters in those days. The telephone existed, obviously, but we used it for practical things. We never telephoned each other to chat; we wrote letters.

MP: I'm convinced that your grandmother was a rare person. You said that you very much wanted to live up to her. Do you feel you have?

McDuff: Yes, I think so. I certainly have done the best that I could.

Mother's Career Frustrations

McDuff: I knew that my mother always felt frustrated in her career. She had great ambition. In *her* mind, I had to ignore the fact that I was a woman and just succeed in my career. She was taken aback when I had my first boyfriend, David, because she thought that would be a hindrance.

MP: When your mother said that you had to ignore the fact that you were a woman, was she suggesting that it was better to be a man?

McDuff: On my mother's side, I come from a line of strong women, starting with my great-grandmother Maud Pember Reeves, who wrote a pioneering book *Round About a Pound a Week* on the London poor. Nevertheless, there was definitely a feeling in my

Figure 12.8 Justin (Blanco White) Waddington, McDuff's mother, on veranda of the house she built in Italy about 1980.

family that men were more important. When I had my daughter, Anna, the very first thing my mother said when I told her was, "Oh, what a pity it's not a boy!" She thought that boys were better than girls. My grandmother didn't have a proper career, and my mother didn't have a full career, though they were both incredibly talented people. They didn't do as much as they could have if they had been men. So, there was definitely a feeling

that to be a woman was to be inferior. My desire not to be second, together with the fact that it is easier to earn a living as a mathematician than as a poet, was why I made the living during my marriage with David; but it did make things difficult, because then I had to do everything. In some ways very self-confident, I also had many feelings of inferiority that took a long time to overcome. There are many contradictions.

MP: *Earlier you said your parents greatly embraced and approved of creativity, and certainly poets are thought to be quite creative. And you referred to David as being a truly brilliant and creative person.*

McDuff: Indeed, indeed, but even so, people are full of contradictions.

MP: *Okay, fair enough. So, she wasn't that pleased with him. Was your father happy with him?*

McDuff: My father didn't pay much attention. He was off in his own world. Anyway, part of the point of trying to live up to my grandmother was to do things that my parents wouldn't approve of. You have to understand that my grandmother didn't behave as the people around her thought she should.

I grew up in the 1960s. In the summer of 1961 I was home alone during the week of the Berlin Wall crisis, and I really believed that a nuclear war would annihilate everyone in a few days. So I joined the Campaign for Nuclear Disarmament. From then on, I too was not willing to be or do everything that people expected of me.

Figure 12.9 Dusa aged about seventeen.

MP: *You have spoken about the importance of getting your father's approval. Did he live long enough to see you achieve professionally at very high levels?*

McDuff: Yes, he did. He died in 1975 when I was a lecturer at York. I heard indirectly that he was very proud of me, at his funeral from a close friend of his. As far as I know, his basic attitude about mathematics was still that it is too dry, but he was much more open to its value as a language for science. Towards the end of his life, very interested in sustainability and environmental issues, he wrote a book and developed an undergraduate course on *Tools for Thought*, how to understand scientific techniques of problem solving. He also organized several meetings on the theme "Towards a Theoretical Biology" together with mathematicians such as Christopher Zeeman and René Thom. But by that time I was under the influence of Gelfand, who thought that Thom's approach was shallow, so my father and I did not even connect on that.

MP: *Well, it would have taken a bit of time before he would have reached a point where he could understand what you had actually done.*

McDuff: I remember one time trying to explain to my mother what I had done. This was when I was a graduate student and I'd had my first idea, proving my first theorem. Despite her willingness to listen and understand, she lost the thread of the ideas as I was going through all the needed definitions—I remember we got lost when we got to "group." It was impossible for her to understand at the level she wanted to without years of training.

MP: *Let's get back to your undergraduate years at Edinburgh. You said that the students who were doing mathematics there didn't seem to be interested in what you really regarded as mathematics. Were they especially utilitarian in their outlook?*

McDuff: You have to understand that I was, you would say, a dreadful snob. I just didn't talk to them, so I don't know what they were interested in. On the other hand, we did have a few recitations together, and if there had been somebody else who was really involved in the classes, then it's conceivable I would have talked to them.

MP: *They didn't strike you as interesting.*

McDuff: They didn't strike me as interesting, and that says a lot about me as well as a lot about them.

MP: *You are referring to students at Edinburgh. What about Cambridge students?*

McDuff: I wasn't at Cambridge for my undergraduate work. I won a scholarship to Cambridge, but I didn't go there. I went to the University of Edinburgh because that's where David was.

MP: *But then you went on to Cambridge for graduate work.*

McDuff: Yes. As a graduate student, I did talk to some of the others, but graduate school in Britain is very different from what it is in America. Much more specialized, students have only three years, though usually with no teaching duties. Because I had distinguished myself at Edinburgh, I was excused from part three of the Cambridge Tripos.

The Tripos is roughly the equivalent of a master's degree; one takes six lecture courses in different subjects, with final exams at the end of the year. If I had done that, then presumably I would have gotten to know both more mathematics and more students. Instead I immediately started working with my advisor G. A. Reid on my dissertation topic, interacting only with the small group of students in functional analysis, none of whom were working on exactly the same subject as I was. After my first semester in Cambridge, I married David, who then came to join me. My life outside mathematics was with him, which cut me off from the other students.

Also, I still was not talking to people. For example, John Conway was there. He was this idiosyncratic person who wore no shoes,

had six kids. . . . I never talked to him. One end of the Common Room was full of his games, which now I think are fun, but in those days I didn't want to be involved. I was terribly serious, seeing mathematics as a very high, abstract, artistic endeavor; games were not part of it.

An Encounter with I. M. Gelfand

MP: *After two years at Cambridge, you went off to Moscow for six months.*

McDuff: Right.

MP: *You ended up working under Gelfand because his was the only name that was familiar to you before you went there. Is that right?*

McDuff: More or less. I hadn't thought that I had to study with anybody. My advisor had not asked: "Well, do you know what you want to do when you get there?" I just hadn't thought about it; I was very naive in many ways. But luckily, when they asked me in the student office at Moscow University who I wanted as my advisor, I said Gelfand, and that turned out to be great.

MP: *Good choice.*

McDuff: His was the first name that came to mind, probably because I had written my undergraduate project on his book on distributions.

MP: *Tell us about getting to know him and the impact that he had on you.*

McDuff: When I first met him, he asked what I was doing, why I was there, and then said that he was much more interested in the fact that my husband David was writing a thesis about the Russian symbolist poet Innokenty Annensky than that I had solved this problem about von Neumann algebras. Nevertheless, he then tried to figure out what he could teach me.

Gelfand amazed me by talking of mathematics as if it were poetry. He tried to explain to me what von Neumann had been trying to do and what the ideas were behind his work. That was a revelation for me—that one could talk about mathematics that way. It is not just some abstract and beautiful construction but is driven by the attempt to understand certain basic phenomena that one tries to capture in some idea or theory. If you can't quite express it one way, you try another. If that doesn't quite work, you try to get further by some completely different approach. There is a whole undercurrent of ideas and questions.

MP: *What is the single biggest thing that he gave you? He was clearly a very inspirational person for you.*

McDuff: He was the first person I had met who saw mathematics in the way that I imagined it. At that point, married to a poet, I was very idealistic. I saw mathematics with its abstract beauty as one way of expressing human thought and creativity. Gelfand also saw mathematics as part of everything else. Whether he was reading books to me, or we were listening to music, for him, that was doing mathematics. At that time in Russia,

Figure 12.10 At Gelfand's seminar, Moscow University fall 1969, McDuff talking, Gelfand explaining.

there were so few outlets for people's creativity that many people became mathematicians who in the West would probably have done other things. We think that mathematical talent is something very special, possessed by only a few. But many different kinds of minds can contribute to mathematics, and they did so in Russia. That's one reason it was such a vibrant and broad mathematical culture. In addition he used so much mathematics, while I knew nothing. He gave me his recent paper to read, "Cohomology of the Lie Algebra of Vector Fields on a Manifold." I had been so narrowly educated that I didn't know what cohomology was, what a Lie algebra was, what a vector field was, or what a manifold was.

MP: *So, he inspired you to learn a lot of new mathematics.*

McDuff: Yes, he opened my eyes to many things, but I was only in Russia for six months.

MP: *A very important six months. It sounds as if he spent a lot of time with you.*

McDuff: Being a young and very inexperienced mathematician from the West, I was a complete novelty. There were very few visitors from the West in those days, and Gelfand was eager to practice his English. He tried to help me find a way forward, suggesting many things for me to read. He once said toward the end of my time in Moscow, when I'd reexpressed one of his results in the language of sheaves that I was just learning: "Oh, you're quick." He realized that I could do some mathematics and gave me a letter of recommendation to help me in the future. He gave me a vision and he spent time with me. He obviously thought I was worthwhile, which was very encouraging.

Gelfand cared deeply about education. In Russia talented young Jewish people couldn't get the education they wanted and were largely shut out of the university system. He found ways to bring mathematics to them, for example through evening schools that he set up.

MP: *He also wrote some very elementary and innovative books on algebra and trigonometry.*

McDuff: He grew up in the provinces, excluded because he was Jewish and poor. He devoted considerable energy to creating opportunities for talented young people, inviting them to his seminar and trying to make the ideas accessible.

Changing Fields—Not Easy

MP: *After those six months in Russia, had your mathematical interests shifted?*

McDuff: When I came back to Cambridge, I was working in a completely different field.

MP: *That falls into what some people in sports would call the guts ball category—shifting fields. At that point you had done work, very good work, in functional analysis.*

McDuff: But I didn't know where to go with it. If I had gone to France and talked to someone like Dixmier, I might have found out what my work in functional analysis was related to and where it might lead; but my thesis advisor didn't really know. I had no clue, and Gelfand was interested in other things. So, what was I to do but try another area of mathematics? It wasn't easy; I was starting again and didn't really have a framework. Gelfand suggested that I work with Frank Adams, but Adams was just in the process of moving to Cambridge and didn't know anything about me. He suggested to me that I study algebraic K-theory, which I did, but it didn't help me get my footing in this new field.

After submitting my PhD upon my return from Russia, I spent two more years in Cambridge as a Science Research Council postdoc, basically learning on my own except that I talked to some topologists. I took a course on four-manifold topology from Casson, went to wonderful lecture courses by Adams on Quillen's work on homotopy groups of spheres, and read a variety of books: Milnor on Morse theory; Lang and Serre on number theory. At that stage, I was still learning passively, not working on any problems.

Since I had no duties, David and I were able to spend several months each spring in my parents' place in Tuscany. This was a peasant cottage, built of stone with no running water or electricity, in the middle of an olive grove owned by Aleksander Zyw, a painter friend of the family. We spent idyllic months there. After each day's work, we would walk up the hill to drink wine with Aleksander and his Scottish wife Leslie and talk about painting, olive trees, and life in general.

Back in Cambridge for the second year of my postdoc, I gave birth to Anna. At the end of that year, the department paid for me to go to a *K*-theory conference at the Battelle Institute in Seattle, where I met Graeme Segal. We got to know each other and started working together. I did the equivalent of a second PhD with him, finally getting back into research by working on problems he suggested. For a long time, feeling totally inadequate because of my ignorance, I had just been trying to learn things with no sense of where the questions are, what I might contribute. If I'd continued on my own, it's not clear to me that I would have found a way back.

MP: *So after your postdoc at Cambridge, you took a position at the University of York. How did that happen?*

McDuff: There were very few jobs in the U.K. that year (1972), just four I believe, and I was lucky enough to get one. Some years

Figure 12.11 McDuff and daughter Anna near Kenilworth, Warks, 1978.

later, my Cambridge advisor said, "What a pity you didn't apply for that lectureship at Cambridge. . . ." Why didn't he suggest it at the time?

I left Cambridge because there were no lectureships that year, and the only possible fellowship was at Girton, where the salary wasn't enough to support a husband and child. I wasn't eligible for one of the better-paying fellowships, say at Trinity, because they were reserved for men.

Although there wasn't overt prejudice at Cambridge when I was a graduate student, the Cambridge system wasn't set up for women to have a career. I was somewhat anomalous, supporting a child and husband, and I couldn't survive on a very small fellowship from Girton. So I left.

MP: *So you went off to York.*

McDuff: York, yes. At that time, I had a nine-month-old child, my first teaching job, and a husband who refused to do anything around the house.

MP: *You were busy coping with lots of responsibilities.*

McDuff: And I was trying to do mathematics, so I was very busy.

MP: *Well, I think a nine-month-old child would be more than enough to occupy your time.*

McDuff: David did look after Anna, I have to say, but he wouldn't change her nappies; I had to drive home five miles every lunchtime for that. He said they were too geometric (they did have to be folded). We also couldn't afford good disposable nappies. I washed them in a machine, and hung them out to dry in the garden.

MP: *My wife and I are greatly enjoying our first grandchild, and I have gotten real insights into how much effort goes into caring for a baby.*

McDuff: It really is a lot of work. It's also hard to do mathematics when your time is chopped up into little pieces.

MIT and Graeme Segal— A Turning Point

MP: *So, after your Russian experience, you came to feel you could do research again.*

McDuff: Slowly, very slowly. I very much enjoyed being at York, teaching for the first time. Feeling less intimidated once out of Cambridge, I organized seminars with other young faculty and learnt with them. We were allowed to carry through some ideas about changing the structure of the undergraduate program, introducing a choice of courses and student projects into the last year. I continued working closely with Segal, mostly via letters, completing a paper on configuration spaces of positive and negative particles.

In 1974–75, I was invited to MIT as a visiting assistant professor. While there I realized how far away I was from being the mathematician I wanted to be, but I also realized that I could do something about it. I became more aware of the relevance of feminist ideas. Before I'd thought that I was beyond all that since I already earned the living, but I've slowly learnt that these matters go much deeper.

In those days, I was still a follower, not interacting with anyone on a basis of equality. I had most inappropriate role models, either sirens such as Lou Andreas Salomé or sufferers such as the Russian poet's wife Nadezhda Mandelstam—neither of much help in becoming a creative mathematician. It was also harder than it is today for a young woman to interact with other (male)

Figure 12.12 McDuff in Toronto, 1974.

mathematicians in a purely professional way as a student or colleague; there were too few of us.

That year, for the first time, I met other female students of mathematics to whom I could relate. I also had a mathematical idea again, the first real idea since my thesis, which grew into a joint paper with Segal on the Group Completion Theorem. (Again we collaborated by mail, since he was in Moscow and Oxford.) The year at MIT was crucial in building up the research side of my career. I woke up and realized that I could affect my life.

MP: *It seems that the influence of Gelfand, Segal, and your MIT experience were key factors in your development as a research mathematician.*

McDuff: Because I had a child, was very busy, and had changed fields, it would have been very easy for me not to succeed as a research mathematician. I read a very interesting book* 20 years ago about women in academia that showed how small but key things determined whether women remained in the academy or were, as they said, "deflected" by advice they received, inconveniently timed moves they had to make to accommodate their husband's career, or an illness. If you're an outsider, it is almost impossible to function. For marginalized people, as women were at that time, it's not only a question of merit but also one of luck. I had talent and perseverance, but I was also lucky.

Broadening Horizons

MP: *Never underestimate the value of luck, but it takes more than that to succeed in research.*

McDuff: For me it's been a long steady haul.

MP: *So, in what was to be year three at York, you went to MIT.*

McDuff: Right.

*Nadya Aisenberg and Mona Harrington, *Women of Academe: Outsiders in the Sacred Grove* (Amherst: University of Massachusetts Press, 1988).

MP: *Did you simply learn of this appointment that was reserved for women and decide to apply?*

McDuff: At that stage I would never have applied; I was invited to come. MIT was looking for women, and I imagine that I. M. Singer had heard about me from Gelfand.

While there, I realized that instead of being envious of other people's opportunities, I could arrange them for myself. I applied to the Institute for Advanced Study and got in. There was a job coming up in Warwick. I didn't have to be at York; I could apply to Warwick. It was that kind of thing.

MP: *It sounds like something of an awakening. So, you went to Warwick for two years.*

McDuff: Yes. I was very happy there.

Figure 12.13 McDuff, about 1985.

MP: *Warwick was a very young university at that point.*

McDuff: York was also a "new university," as they were called. But Warwick had more international connections and possibilities because of the Mathematical Institute set up by Zeeman.

MP: *In the early 1970s, were all Cambridge colleges still single sex?*

McDuff: I felt excluded from Cambridge for that reason. There were one or two other women floating around the department, but they were pretty marginal. The few female graduate students typically got married and left the field after their PhD. There was a senior woman in statistics, but I did not know her. Girton had women mathematicians, Dame Mary Cartwright for example. But although I was formally a member of that college, it had no provision for married students, and I never went there. My one interaction with Mary Cartwright was when I was a schoolgirl applying to Cambridge; I had tea with the Mistress of Girton College, who just happened to be the distinguished mathematician Mary Cartwright. I don't remember knowing that she was a mathematician. She handed me a cup of tea in a delicate cup. It was a formal occasion, not a mathematical one.

MP: *You said that, during this period, your husband David was something of a house dad.*

McDuff: After finishing his PhD, he didn't want a regular job, instead translating poetry from many languages. He was not at all domestic.

MP: *So in 1975, after being at MIT, you went back to York, and that coincided, roughly, with your separation from David.*

McDuff: Right.

MP: *And then on to Warwick in 1976.*

McDuff: That was a very good place for me. I could well have stayed there for the long haul, but I moved to Stony Brook in 1978 for personal reasons I'll talk about later. The two institutions are alike in many ways, not quite at the center of things but rich with opportunities.

Isolation

MP: *You frequently speak of being isolated in your writing.*

McDuff: Right.

MP: *Many people continue to believe that mathematics is a single-person sport and not a team sport. You seem to be suggesting very strongly that interactions with other mathematicians are very important.*

McDuff: I think they are. I talked about this with my husband, Jack Milnor, recently, because he is somebody who almost always works alone; he does talk to people, but not that much. What he said was that when you're learning a subject, it is vital to talk to others. You have to grow up in a community, know where the subject is at, and what the interesting problems are. Once you have a general framework, you can fruitfully work on your own—though even then it's often good to talk to others.

A large part of the problem was my attitude. If I'd had role models of women challenging authority I might have done better—to do research you have to ask questions. I did not know anyone who was attempting to live a similar life, and so it took a long time for things to come together. The late sixties was the time of the "Free University" in Cambridge, lots of far-left politics, very antiauthority. There were very few adults whom I was willing to talk to— Gelfand and Aleksander Zyw—but not my parents or anyone who might have given me sensible advice.

I kept myself apart from most other women since they didn't seem to share my ambitions. I also isolated myself because of my life with David; he was not at all sociable either.

MP: *You said that Gelfand opened a number of fields to you. That might have helped with the isolation problem.*

McDuff: Eventually it did.

MP: *He introduced you to subject matter that he thought you should know and revealed the interconnectedness of many of these ideas. A mathematician I talked to recently said that underscores a defect of American education in mathematics. He was talking about it at the undergraduate level as well as the doctoral level. He claims that students get pushed through a number of required courses and very often finish a bachelor's degree or an advanced degree with only a weak idea* *of how mathematical subjects connect. For that reason he recommends developing new sorts of modern capstone experiences that bring the ideas together. I don't know if that's particularly easy. I think it's easier to do it at the undergraduate level than at the graduate level. Maybe that's part and parcel of education in Russia.*

McDuff: Well, I don't know what education in Russia is like now, but the Gelfand seminar reflected his very broad interests. His attitude was that everything was one.

You can't learn all of mathematics, but the education that I got as a graduate student in Britain was very narrow. Education there is still rather narrow because, at least until very recently, PhD students were allowed only three years. I think it's much better in the States than in Britain. Instead of starting work immediately on some thesis topic, students have one or two years of general courses and can then decide whom to work with. Even though people may still be specialized, they have certainly seen more mathematics when they finish than I had.

MP: *So most new PhDs in Britain are about 24 years old when they complete their degrees.*

McDuff: I was 24 when I'd done mine.

MP: *In the United States, the average is closer to 27, a dramatic difference.*

McDuff: It used to be the case in England that you specialized more as an undergraduate, so you had a bit more mathematics completed before starting graduate school, but not in Edinburgh because the Scottish

university system was more like the American. Of course, if you are in a department with broad enough interests, you can continue developing and growing.

MP: *Do you still feel isolated?*

McDuff: No, I don't feel isolated anymore, but that's fairly recent—really starting when I became interested in symplectic geometry in the mid-1980s.

Symplectic Geometry

MP: *I have to compliment you on your article on an introduction to symplectic geometry. It's the first one that really began to make sense to me. Your exposition is a gift.*

McDuff: Which introduction was that?

MP: *The one you gave at the European Women in Mathematics Conference.*

McDuff: Oh, that recent one.

MP: *As I say, it was a gift.*

McDuff: At first I thought you were talking about the book I wrote with Dietmar Salamon, called *Introduction to Symplectic Geometry*, whose first chapter is "From Classical to Modern." One of the things I did learn in Edinburgh was very old-fashioned classical mechanics, spinning tops, canonical transformations, and all that. Those are the roots of symplectic geometry, so this chapter started off talking about classical mechanics and then discussed Gromov's modern approach to the geometry of Euclidean space.

Jack Milnor

MP: *Somewhere along the line, you encountered Jack Milnor.*

McDuff: I met him when I was at the Institute for Advanced Study in Spring 1976.

MP: *Apparently you and Jack do discuss mathematics.*

McDuff: To some extent. I've never collaborated with him. But we do talk about mathematics, and he occasionally reads things that I write. For example, he read the article of mine you liked, helping me make it understandable to somebody who doesn't know the subject.

MP: *I've heard more than one mathematician say, "Never marry a mathematician in the same field, because that can only lead to conflict." That was one of the reasons for asking that question, but your fields are sufficiently different that mathematical conflicts are less likely to occur.*

McDuff: Well, he's moved out of topology into dynamical systems. There are relations between symplectic geometry and dynamical systems, but we don't talk much about them. He's wonderful when I want to ask questions about topology. We do interact about mathematics, somewhat more when we met than now, but I've always felt that it's better for me to be independent.

MP: *I could have guessed that by now. So, as of the moment, you're continuing to work vigorously in symplectic geometry, not anticipating a move into some other field.*

McDuff: Symplectic geometry is incredibly rich. Many brilliant young people have come into the field. It now encompasses a huge amount of mathematics and relates to many other areas, so I am not tempted to do something completely different. The homotopy theory I studied as a postdoc is relevant, as are many other things I learnt along the way. For example, I first learnt about continued fractions in order to teach a workshop for the Stony Brook undergraduate Women in Science program, but it was a crucial ingredient of my latest work about embedding ellipsoids. Of course, there are many other topics like mirror symmetry and algebraic

Figure 12.14 McDuff and Jack Milnor, about 1978.

geometry that I don't know enough about, but I am more confident that I can learn them as needed.

MP: *There are all kinds of opportunities, even full-blown journals in symplectic geometry.*

McDuff: The *Journal of Symplectic Geometry* is fairly recent, I would say, about five years old or so.

MP: *That's young for a journal.*

McDuff: The field has influence. The ideas of Floer theory, first expressed in symplectic geometry, have now permeated low-dimensional topology, with close connections to gauge theory and homological algebra. Via mirror symmetry, we now understand that symplectic geometry is in some sense the twin of complex geometry.

Figure 12.15 McDuff with son Thomas, 1985.

Dusa McDuff as Role Model

MP: *In 1978 you moved to Stony Brook to be closer to Milnor, who was at the Institute in Princeton, and in 1984 you were married. At Stony Brook you quickly became a full professor and department head. As department head you worked to improve the curriculum. Tell us about that.*

McDuff: I made efforts to improve the undergraduate curriculum at Stony Brook by introducing new classes and working to attract more majors. I like teaching. I was involved in calculus reform, teaching calculus with computers, and experimenting with new curricula. I haven't been able to work

consistently at that because it's just too much to spend a lot of time doing that and have a family and do research. (My son, Thomas, was born in 1984.) At Stony Brook I worked with others to improve the undergraduate program, but there is always a huge amount left to do.

MP: *Like it or not, you've become a role model for women in mathematics. How do you handle that?*

McDuff: At some point I had to decide whether I'd put more effort into doing administration or into improving teaching—getting involved in a major way and doing it properly, or whether I wanted to concentrate on research. There are many prominent women scientists who've recently become deans or university presidents. I decided I didn't want to take that route, since what I really cared about was doing research. That's also important for women, to see that women can excel at research. I recently moved to Barnard partly for that reason, since it gives me a visible platform on which to be both a woman and a research mathematician.

I always try to encourage young women, indeed women at all stages. I had very little advice, and talked to very few people early in my career. Now I talk to people a lot; it's good to encourage dialogue. There are many ways of being a successful woman mathematician. When I gave my acceptance speech after winning the Satter Prize, I said that it would be good for other women to talk about how and why they became mathematicians, for my story is certainly not the only one. A few months later, the *Notices of the American Mathematical Society* published fascinating autobiographical accounts by six other women. That was very valuable because it showed women from different backgrounds, with different motivations and interests, and different ways of contributing.

MP: *Such articles actually do lots of good for mathematics in that they provide a lot of information, counseling as it were, of what it means to be a woman in mathematics and what they're up against.*

McDuff: Now there are several very successful programs designed for women. People have been trying various approaches, some of which have really worked. For example, the Nebraska Conference for Undergraduate Women and the program at the Institute for Advanced Study are both excellent. There are still not enough women in the profession for us to take their presence for granted. Anything that can bring women together and help them meet others, who are possibly quite different but facing some of the same issues, is to my mind very helpful.

MP: *Some women mathematicians I have spoken with believe that employment opportunities for women mathematicians have actually declined in recent years.*

McDuff: It's still very spotty. There are some really brilliant young women mathematicians now; it is much more accepted that women can be good mathematicians.

There's still a lot of hidden prejudice, not just about women. There are so many attributes—accent, color, or background—that make one think a person couldn't possibly be a mathematician. It takes a long time for the academic culture to change.

Some good departments have made serious efforts to overcome this problem. In those that haven't, there aren't many women. In general, I think the situation is much better than it was. There are many more opportunities. Every year there are some women everybody wants to hire. Coming here to MSRI is great fun; there are a lot of young women in the program, which improves the atmosphere for everyone.

One can't be complacent. Everybody should be encouraged, not just women. The

Figure 12.16 McDuff, about 2004.

old attitude used to be that if people are any good they'll survive, and if they don't, there are always others. I disagree; we have to care about everyone.

Career, Family, Husband—
A Lot of Juggling

MP: *In one of the articles you have written you say that you only survived because of the confidence that was instilled in you by the success of your work on von Neumann algebras.*

McDuff: My upbringing also gave me a lot of confidence. I felt I could do everything, and I tried to do everything—to have a career and a family and a brilliant husband.

But it's not possible to keep all those balls up in the air at the same time.

MP: *It's difficult. I don't think it's gotten easier.*

McDuff: It's not gotten much easier, except that attitudes toward gender issues have improved. Marriage is now more of an equal partnership, while in the past women and men were expected to play very different roles. It is less of an anomaly for a young woman to have the ambition to succeed as a mathematician. In my day it was considered so unfeminine that I had to spend a lot of energy proving I was a woman.

At the beginning, before I had found my way as a mathematician, the fact that I had written a good PhD thesis gave me belief in myself. The visibility of my thesis also enabled me to get jobs. No doubt I was known because I was one of the very few women doing research mathematics at that time, but I had also shown that there are infinitely many type II_1 factors, a question left open since the foundational papers of Murray and von Neumann in the 1940s. So that gave me a firm basis on which to build a life.

Thirteen

Donald G. Saari

Deanna Haunsperger

You might expect a member of the National Academy of Sciences, Editor of the *Bulletin of the American Mathematical Society*, author of well over a hundred publications, winner of the Ford, the Allendoerfer, and the Chauvenet Prizes for expository writing, Guggenheim Fellow, and author of the books *The Geometry of Voting* and *Chaotic Elections! A Mathematician Looks at Voting* to be too busy to discuss the importance of mathematics with a group of nonmathematicians or a class of fourth-graders. But if you have read what Donald G. Saari, Distinguished Professor and Director of the Institute for Mathematical Behavioral Sciences at the University of California, Irvine, has written, or if you have had the pleasure of a conversation with this consummate storyteller, you would find you were wrong. Proselytizing the beauty, power, and ubiquity of mathematics is not an obligation to Saari—it's his passion. One afternoon at Mathfest 2003 I had the opportunity to talk with him about growing up in Michigan, his education, and his mathematics.

"But Who Needed Money?"

MP: *Tell me about growing up and becoming a mathematician.*

Saari: I grew up in the Upper Peninsula of Michigan, in a Finnish-American region right in the midst of the copper-mining area. It was like a frontier town, where copper mining was the big industry. We lived on the edge of town, so as a kid I spent quite a bit of time in the woods, hunting, camping, snowshoeing, fishing, and even climbing down and exploring abandoned copper mines for mineral samples. I was involved in Boy Scouts, in athletics, in forensics, in class plays. You name it, I was involved in it.

My parents were idealists. My mother graduated as valedictorian of her class at sixteen; by the time she was seventeen, the people in the local churches were campaigning against her because she believed in, of all things, women's rights. Things that today we would accept as absolutely trivial were not taken lightly at that time in some of the

Figure 13.1 Even as a young boy, Saari was outstanding in his field.

conservative areas. My father also was very much of an idealist. He spent time in jail for leading a protest against the fact that farmers were losing their farms during the Great Depression. Together they decided to channel their ideas and energy in the labor movement. They were organizers and community activists in the area, which meant there was a lot of idealism in our family, many inspiring discussions, but no money. Never any money. But who needed money?

Grade school and high school were very, very delightful. Perhaps I enjoyed people a little too much, as I would gab and joke even during class time. As a result I served more than my share of detentions. The teacher in charge of detentions was Bill Brotherton, our algebra teacher. He was always delighted when he heard that I was headed for the after-school detention session because he would bring in all sorts of math books. I was essentially getting a free math tutorial on

different aspects of math that weren't taught in my school at that time. He was very good, very instrumental in developing my interest in mathematics.

I don't recall what I planned to do when I grew up. I always enjoyed academics, but I enjoyed athletics, I enjoyed acting, I enjoyed interacting with people—quite frankly, I always enjoy a good time and bad jokes. I knew I liked mathematics, but at that time I didn't realize you could have a good living, a good life in mathematics. At that time I thought that the only thing you could do with mathematics is be a high school teacher. While being a high school teacher would have been fine, I just wanted to move out of the area and look around.

When I graduated from high school I applied to two schools: one of the Ivies and Michigan Tech. I was accepted to both. I couldn't afford to go to the Ivy; I couldn't even afford to pay the transportation to get there, and Michigan Tech was on the other side of town. The scholarship they gave me paid all my university expenses, so I went to Michigan Tech for financial reasons, but it was an excellent experience. Absolutely. In our graduating class we had George Gasper who's now at Northwestern, Dan Maki who's at Indiana, Jim Thomas at Colorado State, and a couple of others. I'd say at least half of us ended up in university professorships.

A Triple Major?

When I was at Michigan Tech, I had a triple major: social life, athletics, and campus politics. I entered as a typical undergraduate,

Figure 13.2 At Michigan Tech in the 1960s, there was always time for a photo op.

changing majors as fast as some people change shirts. I started off in chemistry primarily because I had very high grades in the subject, but I hated the labs. Then I changed into electrical engineering because I heard it was one of the more difficult majors, but I was responsible for a couple of explosions in the labs there, too, trying to hurry things through to go out to have coffee. I still recall my instructor, putting his arm around my shoulder and hopefully recommending, "Don, have you considered a nonlab major?" So, I moved into math because I enjoyed it so much, and I found it easy. Gene Ortner was my instructor in my first course in abstract algebra. Oh, was that delightful; the way he taught it made that subject seem so beautiful. And my major just evolved into mathematics because that was my real interest anyway.

Always has been, always will be. It was that abstract algebra course, a couple of the other courses, and the dedicated Tech instructors at the time who motivated me to go on to graduate school: involved instructors can have a huge impact on the future life of a student!

Harry Pollard

I was accepted to several graduate schools, but quite frankly I was not very sophisticated at making the decision about where to go. As I looked through the places where I applied and the offers I had, I thought, well Purdue looks pretty good, so I went to Purdue. I'm pleased with my decision. I had an absolutely tremendous experience at Purdue. I was one of those graduate students who, when I arrived, because of my positive experience in algebra, knew I was going to be an algebraist. Absolutely, there was no question; I was going to be an algebraist. I loved the courses I took, but then I discovered algebraic topology, and is that a beautiful subject! So then I was going to be an algebraic topologist. After that I got hooked by analysis. In complex analysis the powerful results you learn at an early stage are just so beautiful, so I decided to become an analyst. Then I discovered applied mathematics, and that was nice, but I realized I had to understand functional analysis better if I'm going to do partial differential equations. So I started toward a thesis in functional analysis when my advisor said "Don, I know you like teaching. Harry Pollard is considered our best teacher. You won't like the material (celestial mechanics, the n-body problem), but why don't you sit in so you can see how he teaches?" Being a student, I made the obvious bad jokes about how celestial mechanics is the study of how heavenly bodies move, but I agreed, and once I was in the class, I found the material so fascinating and so nice, that I made my next switch to the n-body problem, and I wrote my thesis about collisions in the n-body problem.

Harry Pollard was different. Delightful. We would get together and talk about mathematics maybe once or twice a week. Any branch of mathematics. He never posed a research problem for me or anything else; we would just sit down and talk about mathematics. It was enjoyable and educational, but Harry had no idea what I was doing or if I was even working. The first he really learned about my research was when I handed him a draft of my thesis. He responded, "Oh, so you are working!"

I was at Purdue five years, but one year was spent courting my wife Lil. Of course you understand, when I finally got enough courage to ask her to marry me, her response was "What? And be 'Saari' for the rest of my life?!"

Northwestern

After Purdue, I went to Yale. I was there only a short time when the phone rang. It was Ralph Boas from Northwestern, and he said, "I understand you're looking for a job." Well, I wasn't, but I said "Sure!" So he invited me out and I gave a "job-talk" lecture on the collision orbits in the Newtonian n-body problem. Some of the techniques I was using in the n-body problem, both in the work on collision orbits and in my later work describing the evolution of the n-body problem, made use of an analytic tool called nonlinear Tauberian theorems. They should be called Hardy-Littlewood theorems because they're the ones who extracted them from theorems that had been done earlier by Tauber. To give a little history, A. F. Wintner at Johns Hopkins University in the early thirties asked Ralph Boas to look at a result by this Finnish mathematician by the name of K. F. Sundman. Wintner suggested that there was a Tauberian argument hidden in Sundman's analytic argument about two-body collision orbits. Ralph worked on it and, indeed, extracted a Tauberian theorem out of it. In separate papers, Harry and I generalized Ralph's insights into a nonlinear Tauberian theorem that became one of the tools I used extensively in my work

It is encouraging that there are many amateur mathematicians, but at times this can be a serious time drain. In the 1970s, during the Cold War, a short guy in a sports coat and claiming to be an engineer asked if he could talk with me. When he wanted to close my door, I became suspicious. "Why?" I asked. His response, "I don't want the Russians to know what I am going to say."

Standing by my blackboard with a proud smile, he made his big announcement, "What would you say if I told you that I solved Fermat's Last Theorem?" Thinking quickly, I responded: "Nothing. This is because our department chair divides up all of the unsolved problems among the faculty, and I was not assigned that problem. Therefore I'm not allowed to even look at it." After I answered his obvious next question by identifying "who had been assigned Fermat's Last Problem," my colleague Len Evans spent the next three days first finding the error in his proof, and then convincing him that there was an error.

on the Newtonian n-body problem. Ralph was at Northwestern, and I was an admirer of his work, as well as that of Avner Friedman, Bob Williams, and a large number of other people, so there were excellent reasons for me to be interested in a job at Northwestern. When they offered me a position, I took it instantly; we moved to Evanston, where we remained for thirty-two years.

MP: *You have a wide variety of mathematical interests. You're published in celestial mechanics, dynamical systems, social choice, mathematical economics, game theory, voting theory, and mathematical psychology. How did you get interested in so many things? What attracts you to a problem?*

Saari: I was in my first year at Northwestern teaching a graduate-level course in functional analysis. I was asked to let my colleagues know who our best students were. When I said Jim Jordan, they said, "He's not one of ours." So I identified the second best by name. "Well, he's not one of ours either." Well, the third best is so-and-so. "She's not a math major either." Something strange was going on, so I asked these students, "What are you doing in here beating up on our poor math majors?" They were graduate students in economics. At that time I probably thought that economists looked at the stock market or something; I discovered that when they examined economies with a large number of commodities or economies with a large number of agents, in a natural manner the issues could be posed as problems in functional analysis. I became intrigued by what they were doing. A couple of years later, John Ledyard in the economics department and I ran a seminar on the infinite-dimensional economy where I would worry about the mathematics and he would worry about the economics. The more I learned, the more it became clear that there were some absolutely fascinating issues over there.

Another source of my interest came from Hugo Sonnenschein, who left Northwestern for Princeton, then became provost at Princeton, then president of the University of Chicago. Hugo proved the very interesting theorem that in the usual Adam Smith story of supply and demand, rather than behaving nicely, the economy could do anything.

In modern terms, this meant that the supply and demand story could be as chaotic as you'd like it to be. This was difficult to believe. So I became intrigued, trying to figure out if it could be generalized. My generalization proves that this is a very robust result.

What Makes a Problem Interesting?

I work on whatever intrigues me at the moment. I'm writing a book right now on the n-body problem with some new results. I am just finishing up a paper on mathematical psychology on how individuals make decisions. The work of Duncan Luce has been very important in this area. He received his PhD in mathematics from MIT. Then he and Howard Raiffa wrote a book that had, and still has, a big impact on game theory. Next Duncan became interested in mathematical psychology, and he is one of the founders of this field. I started reading some of Luce's work and found that it was beautiful. My paper in this area extends some of his work. Quite frankly, the social sciences are becoming more mathematically sophisticated. If you attend a mathematical economics conference, you'll hear talks where they're using various aspects of topology or functional analysis, and they're very comfortable with all of these different concepts. This diversity is not true with the other social sciences, but even these other areas are becoming more mathematically sophisticated as time goes on. Attending these conferences, you begin to discover some excellent problems and

issues. While the social scientists are very good, they don't always understand all of the mathematics or the applications. This is where mathematicians can make a tremendous contribution to these areas.

For me a problem is interesting if you really don't understand the underlying structure—where there are no clues about what's going on. For example, when I started working on the evolution of the n-body problem, we knew Newton's two-body solution and the work of the French mathematician Jean Chazy in the 1920s for the three-body problem. But after that there was no clue, no idea, what would happen in general for the four-body problem, the five-body problem, the n-body problem. The same was true with collisions. We knew what happened if two particles collide or three particles collide, but multiple collisions, or collisions that happened with a five-body collision over here and a three-body collision over there?! How likely are they? Since there was no known structure, no guidelines, I just found the challenge to be absolutely irresistible.

In voting theory, only a couple of paradoxical results were known about what could happen in tallying elections. I thought there should be a way to find all possible paradoxes. So I became intrigued by that question. It is interesting how, by borrowing and modifying notions from chaotic dynamics, I was able to characterize all possible paradoxes that could ever occur. Then, to explain why all of them do occur, I used orbits of symmetry groups.

What doesn't interest me is when the overall structure of a problem is basically known, and it's primarily technical details that need to be worked out. Once the overall structure is understood, I lose interest in the problem. Sometimes I lose interest so much in the problem it takes me a couple of years to get around to writing up results, if ever! Bad strategy because often there are many interesting results just waiting to be extracted once the structure is understood. For instance, I have been lecturing on a result in qualitative evolutionary game theory for over three years, but I have yet to write it up.

Figure 13.3 Saari with his wife Lillian in Brazil in the 1980s.

Successful Mathematicians

MP: *What do you think makes a mathematician successful?*

Saari: I really don't know. Hard work—after all, mathematics is difficult. And you have to enjoy the mathematics. You have to say "I can't wait until I start again tomorrow." I'm not interested in a problem because it is known or technically difficult; for me it has to be important for our understanding of a field. But for me, I think what creates this drive to do mathematics is curiosity. You just plain must be curious. But curiosity becomes almost a demon. It can take over what's going on in your life. Earlier I joked about when I proposed to my wife; let me add here that she is very important to me. I think for any successful mathematician, the spouse deserves considerable credit. Just imagine: you're sitting down in the middle of a nice, romantic dinner—and then, instead of sweet nothings, a spaced-out look comes over

Figure 13.4 Saari with daughters Anneli and Katri in the 1970s outside their Evanston home.

you. That's because all of a sudden you see the right mathematical relationship that has been bugging you for the last few months. Wow! With something like that, the only way to keep the marriage strong is to have an incredibly understanding wife! I try to be polite when this happens and snap out of it, but I do scribble on the napkins.

MP: *What of your mathematical work do you like best?*

Saari: I really like my work on the Newtonian *n*-body problem—the work on collisions, the work on the evolution, the work on restrictions on motion, various things like that. The reason I like my work in that direction is that it was very different from what had been done at that time, and it speaks to what I found to be central issues of how the universe evolves, and so forth. I'm proud of my work in voting theory; again, it was a very different direction from what had been done at that time; it answered a lot of questions. What I also liked is that my work goes beyond the traditional approach of finding negative results (such as stating what is flawed or impossible) to show how to find positive results, such as a new interpretation of Arrow's impossibility theorem showing that it does not mean what Arrow thought it meant. My work in mathematical economics I like because it finally convinced me that certain assertions I did not previously believe really were true. Also, this work got me back to nonlinear functional analysis and foliation theory, and so the mathematics is just fun.

UC Irvine

MP: *It sounds as if you could never be happy in just one field.*

Saari: Probably not; in fact, that's one of the reasons that I moved to the University of California, Irvine. I met Duncan Luce, whom I mentioned earlier. Duncan invited me to come out and spend a quarter. Well, I thought I'd go out and see what's going on, learn some of the new mathematical ideas that were being developed there, and then return to Northwestern.

Well, I ran into many excellent people, and they started recruiting me. My initial response was that I wasn't interested. I mean Northwestern has always been very, very good to me, and I had a tremendous time at Northwestern. My ties to Northwestern were so tight and so strong, and I had such a positive experience there, that the idea of leaving was totally out of the question. My daughter Anneli and her husband were thinking about buying a home near us in Evanston, so they asked, "Dad, I know that every so often schools are recruiting you. Are you going to leave?" I answered, "No, Mom and I will be here until hell freezes over." Two months later I called her to ask, "Have you heard the weather report? It's snowing in hell." I decided that if I'm going to be intellectually honest with myself that I am attracted by where there are ideas that I find intriguing, I would have to go. I would have to leave Northwestern and move to UCI. It was hard leaving Northwestern because, again, I made so many close friends and have such close ties. But I'm glad we moved; it's been very exciting.

MP: *You have thirteen mathematical children in the United States and several de facto in France. What do you think makes a successful graduate advisor?*

Saari: Encouraging, giving confidence to the student, and letting the student develop what he or she thinks is important. I think that a way to crush a graduate student and his

Figure 13.5 In the Bird's Nest in Beijing (L to R) with his academic grandson, academic daughter, Saari, wife Lillian, and academic son.

or her career is to dictate the problem and outline what has to be done. This is because, really, a main role of a graduate advisor is to help students build self-confidence. The advisor should help them understand the serious and important issues of the field, help them learn how to select problems, and then give them encouragement and let them develop. With different students, I have different ways of working. With some, I talk with them only occasionally to find out what they are doing. But there are others who I make come in at 9 o'clock in the morning to tell me what they did the night before, then again at 5 o'clock in the afternoon to tell me what they've done since 9 o'clock that morning! With different students, different styles: whatever it takes to help the individual student develop. I really dislike the idea of giving a problem because if you don't learn how to find a problem, when you graduate you'll find it very difficult to write that next

paper beyond your thesis. There are so many people who just write up their thesis and are not sure what to do next. A very important part of the apprenticeship is to help the student learn how to find a problem, ask good questions, and not become discouraged when it takes several tries to get any results.

Exposition

MP: *How do you help them mature as teachers?*

Saari: I think that good mathematical research is mathematics that influences the way people in a particular research area think. And good teaching has very little to do with good penmanship at the blackboard; it's whatever it takes to influence the way students think. Also, I'm a very strong believer in the importance of exposition. After all, what we do in mathematics is very important, but most people don't know that. The key to exposition is to influence the way people think. Aha, notice how that same phrase keeps coming up? While I was at Northwestern I created a course for the first-year graduate students on how to teach. Again, the course was pretty much self-discovery. Each student would give a lecture, and the other students would critique it. Then we would talk in terms of what was effective and what was not, but the key theme at all times was that your job in the classroom is to influence the way students think about the subject material. Doing mathematics, teaching, exposition, they all should be fun. If you look at it in terms of influencing the way people think about issues, then it is fun.

Figure 13.6 Zip-lining through the trees, about 70 feet above the ground, in Mexico, December 2008.

MP: *Is that what inspires you in your Editorship of the* Bulletin *as well?*

Saari: What is the purpose of the *Bulletin*? It's not to publish articles for a small select community. If you're writing a paper in number theory, or dynamical systems, or some other area, and if that article is written in a manner that only experts in that area can read, we're doing a disservice to more than 27,000 readers. For the *Bulletin* to play the role that the AMS believes it should play in terms of exposition, most readers should be able to read most articles, at least partway through. I must confess that we have not reached that point yet, but we've been working very hard. Again, the theme is that the *Bulletin* articles should influence how the general mathematical community thinks about mathematics.

MP: *It seems that you've had a little mathematical influence in your home as your daughters both have mathematical abilities.*

Saari: Oh yes, both my daughters do. Anneli works in mathematical instruction with the emotionally, behaviorally, and mentally challenged; she teaches them mathematics all the way from elementary notions through calculus. Right now she's in the public schools. In a hospital she worked in, she was telling me about some of the abilities of the emotionally challenged. They may be emotionally troubled, but that doesn't mean they're not smart. She was teaching one of the boys, who was in eighth or ninth grade, calculus, and he was really worrying about the proofs! She and I wrote an article about the mathematics of psychology a few years back; the exchange was delightful.

When my older daughter Katri was in graduate school, she found some data about the propositions for the state of California, did the statistical analysis, and found that something like millions upon millions of voters were voting on about twenty propositions, yet not one of the marked ballots agreed with the election results. She conjectured that that's not unusual even with a wide choice of probabilities. She had some ideas of how to prove it, and I thought it should be done a different way. So we started talking back and forth, and we proved a couple of theorems. Much fun!

Semi-values

Then we did a second paper about semi-values. It all started when she asked me what I thought about them. My answer was not much because I didn't even know what they were. When she started explaining them to me, we found some anomalies, some paradoxes that could occur, but these conclusions were isolated. We thought we should be able to prove a theorem that classifies, catches all possible paradoxes that ever could occur with semi-values of a certain class, or power indices. To explain these terms, power indices are used when you are trying to measure political power. For example, in the Senate when you have fifty-one Republicans and forty-nine Democrats, you may think they have essentially the same power. But that's not the case because Republicans can be dictatorial; they have almost complete power in the Senate because, with the voting rules, they can always win. Or another case would be in the Electoral College: How

Figure 13.7 Saari in line, with sword in hand, to receive an honorary doctorate from the University of Turku in Finland, 2009.

much power does the state of Wyoming have over California in the Electoral College when they're trying to elect a president? The issue is, how often can a group make the crucial difference? But there are all sorts of different power indices: the Shapley value, the Banzhof value, and on and on and on, and they can give different answers. So we started looking at the issue to try to find a general theorem that would compare all of these different indices to find everything that could ever possibly go wrong with them. We succeeded, and we had a lot of fun doing that. Working with her was educational for me. What I also liked was that this co-author would say, "I don't think you understand what you're talking about!" This co-author could be tough!

MP: *What does your wife think of your mathematics?*

Saari: Lillian has played an important role in my career; she's been incredibly supportive. Mathematics is a blend between creativity and high-powered muscle power, technical strength. There is no way that you're going to be able to describe adequately the technical aspects to a nonmathematician. But creativity involves concepts, and concepts translate across different disciplines. Consequently, it should be possible with most ideas to describe them to someone who's not in the field. Okay, to do so, you have to learn how to tell a good, maybe wild, story that captures the basic idea in everyday terms. The process of learning how to describe an idea in this manner is not easy, but doing so has important advantages because it forces you to rethink through the concepts, to separate the concepts from the technical aspects. In that respect, Lil is my favorite and valued co-author. We talk about all the different type things that each of us does. For instance, in her work when she was still a teacher, we would talk about the problems she had. She is a very important behind-the-scenes co-author.

Lil's very patient. Of course, occasionally I get that glazed look all of a sudden, and she says "Oh, cripe, here we go again." Or when

Figure 13.8 Taking a break from theorem proving, Don and Lillian Saari in Paris.

I promise I'm going to be home by 5:00, and 7:30 comes and goes and the dinner is cold, and she calls down to the office to ask "Are you still at work?" She knows precisely that that is where I am.

MP: *What have been some moments that have stood out for you in your career so far?*

Saari: What are some of the highlights? Oh, after you prove a theorem that you've been working on for two or three years. Let's face it. I don't know if it's a high or it's finally "ahhhh," exhaustion! But it's really delightful. Some moments that stand out are when graduate students make a nice discovery. You can see a graduate student grow into a mathematician, from a student into a professional. Those are really delightful experiences. Working with other students, with colleagues, with other people in the field, it's fun watching people do well. I enjoyed working with the general public in trying to get them excited about the general importance and what is mathematics.

I've been interviewed a couple of times on television and radio for stories about mathematics, given lectures to CEOs about why mathematics is important to them, talked to fourth graders or high school classes, colleges, public lectures (where of course I start off, "Let X be a nonseparable Banach space . . ."), Rotary clubs, and so forth.

It's very important for the general community to appreciate the power and the importance of mathematics. I think that most people understand the power somewhat, but if they have a certain awe and view mathematics as something that is just abstract, they begin to view it as something like philosophy. While I enjoy philosophy, I most surely don't want mathematicians equated with philosophers—particularly at salary and budget time! Mathematics plays a crucial role in the driving of science, the physical sciences, the social sciences, engineering, and so on. But most people are completely unaware of that, including people in the sciences. The central role of mathematics has to be known to our legislators both within the state and the United States, to our administrators at our local universities, to the general public, which supports a large number of these things. We may never reach the point where there is a total understanding and acceptance, but the more inroads we can make, the better. In France they occasionally have articles at the layperson level in the newspapers that are written by prominent mathematicians. I suspect that for the most part the French population has a better

Figure 13.9 Saari lecturing in Japan.

understanding of what mathematics is and the importance of it. We should do more of that.

I do enjoy when I'm asked by someone what mathematics is good for. I make them tell me what they find particularly important, and ask "Where did this come from?" Then we start refining the underlining structure or ideas, and very quickly you can narrow it down to where mathematics is of crucial importance, or where mathematics can make central contributions. I'm a strong believer that interesting mathematics can be found at the heart of almost anything.

Fourteen

Atle Selberg

Gerald L. Alexanderson

Atle Selberg was one of the most distinguished number theorists of the twentieth century, having contributed over a long career to our understanding of the primes, through his work in analytic number theory, specifically his results on the Riemann zeta function and on sieves. In 1949 he produced an elementary, but by no means simple, proof of the Prime Number Theorem, which had previously required deep theorems from complex analysis. The mathematical literature abounds in references to his work—the Selberg sieve, the Selberg trace formula, the Selberg zeta function, the Selberg identity, the Selberg asymptotic formula, and so on.

At the International Congress of Mathematicians held in Cambridge, Massachusetts in 1950, he received one of the two Fields Medals given that year. Only two had been awarded previously, at the Congress in Oslo in 1936. So the importance of his work was recognized early on. He received his PhD at the University of Oslo in 1943 and was at the Institute for Advanced Study at Princeton continuously from 1949 until his death. In 1986 he won the prestigious Wolf Prize.

The following conversation was held in the offices of the American Institute of Mathematics in Palo Alto, California, in June 1999. A draft of the transcription was found in Professor Selberg's papers after his death. We are grateful to Professors Dennis Hejhal and Peter Sarnak, both long-time colleagues of Professor Selberg, for their help in clarifying some elements of the text and to Professor Selberg's family—his wife, Betty Compton Selberg, son Lars, and daughter Ingrid—for their help in filling a few gaps in the details of Selberg's early years in Norway, as well as clarifying some stylistic infelicities in the earlier draft.

"I Essentially Had to Find My Way Entirely Alone"

MP: *You have been at the Institute for Advanced Study for a long time, over fifty years, I believe.*

Selberg: Well, I'll tell you. I came in August of 1947 as a temporary member and I stayed for a year. Then I took a position at Syracuse. I thought it might be interesting to have the experience of an American university. While I was there I got an offer to come back to the Institute, so I returned there in August 1949. Come late August of this year, I will have been there continuously for fifty years.

Figure 14.1 A photo of Selberg from a Norwegian newspaper in 1949, the year in which he was made a permanent member of the IAS.

MP: *So what are you now, a Norwegian mathematician, an American mathematician, a Norwegian-American mathematician?*

Selberg: I think that in many ways I would characterize myself as a Norwegian mathematician more than an American one because I have a very different background. In a sense I found my way in an environment where I essentially had to find my way entirely alone, because even when I came to Oslo, there was no professor there who had interests similar to those I already had. The closest one in Norway would have been Viggo Brun, but he was at the Norwegian Institute of Technology in Trondheim. But even had he been in Oslo, I always had difficulties understanding the way he thought. I had tried even before I came to Oslo to read some articles that he had written about

the sieve method. They had some interesting geometrical diagrams and figures and so forth, but I looked at them repeatedly, and they didn't mean anything to me. I only understood about the sieve method much later when I got into it from a different side, when I first invented my own sieve method and reflected on the principle behind it. And then finally I understood his articles.

MP: *Just Wednesday of this week, using Viggo Brun's sieve, I finished for my undergraduate class in number theory the proof of the convergence of the sum of the reciprocals of the twin primes. And it took me three days to get through the argument. For undergraduates it's a bit of a challenge.*

Selberg: But you probably followed the way that Hans Rademacher did it.

MP: *Exactly.*

Selberg: You see, until Rademacher did this, I think hardly anyone understood it.

MP: *I have never gone back to look at the original—and maybe it's just as well.*

Selberg: Brun had a very original mind. And he was a very nice man. I was very fond of him as a human being.

MP: *How long did he live? I can't remember.*

Selberg: He died in 1978. I did a memorial speech for him at the Academy of Science and Letters in Oslo, in '79. He was 93 years old. He was born in 1885. It was a good year for mathematicians. Hermann Weyl was born that year, and J. E. Littlewood. And there were probably some others.

MP: *A very good year. I was a student of George Pólya here at Stanford, and I was thinking about something just the other day. I saw the film about S. Ramanujan—I was showing it to my number theory class, actually—and it occurred to me that Pólya and Ramanujan were both born in 1887 (thus making it not a bad year either!). It occurred to me that Pólya went to Cambridge and Oxford in 1924 to work with G. H. Hardy and Littlewood as a "young man." But by the time he got to Cambridge, Ramanujan had been dead for four years. I knew Pólya for the last thirty years of his life so, had Ramanujan lived, I could have encountered Ramanujan at some point. But because he died so young, he seems to be someone from way back in history somewhere.*

Selberg: That is true. Well, of course, Pólya was a very important figure in many ways. And he might have met Ramanujan elsewhere. Ramanujan was very important for me. I came across him by accident, in a way. There was a professor of mathematics in Oslo, Carl Størmer, who had started out in pure mathematics and had some results in number theory going way back. The results still carry his name, in relation to solutions of the Pell equation. Later he mostly investigated the Northern Lights. He still kept up an interest in number theory, and, probably spurred on by the appearance of Hardy's book about Ramanujan, he wrote an article about Ramanujan in the periodical, the *Nordisk Matematisk Tidskrift* ("Tidskrift"

is a literal translation of Zeitschrift—you don't say that in English). This is a magazine directed mostly to teachers in the gymnasium and to students who have an interest in mathematics, also to some amateurs. In this article he quoted a number of Ramanujan's results and wrote about his life. I saw it and found it extremely fascinating. I had been reading mathematics on my own; at that time I was probably sixteen.

Somewhat later one of my brothers [Sigmund], who was still at the University and interested in mathematics, already interested in number theory (diophantine equations), brought home at vacation time Ramanujan's *Collected Works*, which he had taken out of the University Library. So I had an occasion to see that, which was very important. I did have some interest in number theory before, and that was also occasioned by that same brother who had pointed out to me in a book in my father's library—my father had a quite large mathematical library—an algebra book by a Frenchman, J. A. Serret, which had a chapter about P. L. Chebyshev's work on primes. It was written before the work by Jacques Hadamard and Charles de la Vallée-Poussin, but it had Chebyshev's inequalities and such. So I read that, and it fascinated me very much. I must say I wasn't interested in much of anything in the rest of that book.

MP: *Yes, I've seen the book, and it is certainly not uniformly interesting.*

Selberg: So I think these two things pretty much gave me direction. There was a third

Figure 14.2 A happy lecturer—Selberg at Columbia University

thing. In the middle 1930s there came the work of Erich Hecke in connection with the Hecke operators, bringing together the functional properties that lie behind the Dirichlet series and tying this up with the multiplicative properties. That I found very fascinating. I thought at the time that this would give access to things like the Riemann hypothesis. Things turned out not to be quite that simple. And they still haven't become simple.

Growing Up in Norway

MP: *No, not at all. You mentioned that your father had a really good mathematical library. Was he a mathematician?*

Selberg: Actually, not originally, but he became one. He was the youngest boy on a farm, and he stayed home after the older boys had left. He left when he was about twenty—before that he had only elementary schooling that he got in the countryside—and he tried to educate himself by taking some quick courses that led him to the final exam of the gymnasium in one year. Then he took some teacher training and worked as a teacher for a couple of years. After that he moved into banking for a while and later into life insurance, where he did very well. In 1920–1921 he was the head of the western Norway division of the largest life insurance company. But he had become very interested in mathematics. He didn't like the work he was in and wanted to go back into education, so he went to the university to take exams from time to time. Mathematics was his favorite field. He finished the necessary education to become a teacher in the gymnasium with mathematics his main specialty. Later he got a doctor's degree. He was then about fifty years old, with nine children.

MP: *Nine children!*

Selberg: I am the youngest. I had four brothers and four sisters. Now I have only one sister left.

Because he was not in a place where there was a good mathematical library, he had to buy the books for himself. He bought a lot of good things. His main interest was algebra, but he had lots of other stuff as well, function theory, collected works (Riemann's, for example, and Abel's, of course, and Gauss's), and he also subscribed to Sophus Lie's collected works that were coming out at that time, also the *Enzyklopädie der Mathematischen Wissenschaften*. Of course, as the years went by that took up a very large space! He ended up with a quite large library. He

never tried to influence me to go into mathematics, so I picked it up on my own.

MP: *Didn't you have a second brother who was a mathematician?*

Selberg: Well, I had my older brother Henrik, eleven years older than I, who actually worked in value distribution theory, in particular, for the so-called algebraic functions, those that are finitely many valued, also in statistics. He worked some in industry for the Nobel Company, which was mainly interested in explosions and detonations. He was very helpful to me when I came to the university; he was then already on the university staff, not as a full professor but something corresponding to an associate professor. And actually when I wrote my first paper in the late summer of 1935, the year I was coming to the university, he typed it out for me. And beyond that, because of my bad handwriting, I think, he then wrote in the formulas. At that time you couldn't really do many formulas by typewriter; you wrote them in by hand. So he was very helpful in those ways, but we did not have the overlapping interests that my other brother and I had.

MP: *When the family was growing up, where were you? You were not in Oslo, right?*

Selberg: No, actually, the first place I can remember is called Voss, which is on the line between Bergen and Oslo, on the western side of the mountains. It's the most substantial station before you get to Bergen. I was not born there; I was born in the South of Norway. I have no recollections of the place, and I've never even gone back to look at it, though it had some meaning for my older siblings. But I was living outside Bergen when I got my early schooling, elementary school. For middle school I went to Bergen by train. It was rather nice in a way, sort of a commuter train that took a little over half an hour. We did a lot of school work on the train. There were quite a few students commuting like that. Then the family moved to the eastern part. My father was really working in the insurance business. In 1930–1931 he was appointed principal (or rector as it was called then) in the school system in Gjøvik, in the gymnasium, which he held for a number of years. At the same time he had another position as inspector of the lower schools. This is something that was not usually combined, but I think it gave him a slightly higher salary, which he may have needed with all his children. He had taken a very serious loss in income when he left the insurance company. There had been a boom right around the time of the First World War, and then there came a downturn, particularly in shipping, and that affected Norway. I think he may not have realized how severe the loss in income might be.

Interested in Numbers at an Early Age

At any rate I became interested in mathematics without anyone in the family pushing me. I was, at an early age, somewhat interested in numbers. I remember I was out with some other boys playing a ball game, not soccer, but another ball game that was played particularly by children, somewhat related to the American baseball. It's also somewhat different and is played with a much softer ball. And the bat is also somewhat different. But it had many of the features of baseball—bases and running and such. And during these games there are times when you stand around and don't have much to do. I remember that by doing some mental calculations in my head I noticed that differences between the consecutive squares give you the consecutive odd numbers. I did it just by numerical examples; I didn't know how to use letters. I found sort of a mental proof by interjecting between n^2 and $(n + 1)^2$ the number $n(n + 1)$. Then you can easily find the difference on both sides. It made quite an impression on me; it was the first time I made a mathematical discovery.

MP: *And this was at age . . .*

Selberg: I think about ten. Later I started looking at some of the books in my father's library. The thing that really started me reading mathematics seriously was a book that included Leibniz's formula for $\pi/4 = 1 - 1/3 + 1/5$, and so forth. That seemed a very striking statement. So to find out how all of this hangs together, I decided I wanted to read that book. It was a set of lectures by Professor Størmer, whom I mentioned earlier. It was a large course in calculus with a little bit of complex variables. This was a rather strange thing to start with. The book started by defining the real numbers with the Dedekind cuts. It's a wonder I didn't give up right there, because I found it very boring. I couldn't for the life of me see what was the use of it. I didn't see any difficulty with the concept of a real number. I thought at that age that you can write anything down as an infinite expansion, and that gives you a number.

MP: So much for too much rigor too early!

Euler

Selberg: Yes. Take Euler. I think he thought that he knew perfectly well what a real number was. He could calculate with them; he could do anything. I don't think he was in any danger of making mathematical errors because he didn't understand properly what a real number was.

MP: I think I'm beginning to suspect that you would not have seen a future for yourself in foundations of mathematics.

Selberg: No! Certainly not. I was very happy when I finished with that chapter and then it started to become interesting. It was a very lopsided mathematical education, you see, because it meant that I met the trigonometric functions first in terms of the Euler formula, in terms of the exponentials. I had trigonometry in school much, much later.

MP: It's interesting that you cite Euler. Some of my favorite mathematicians, when asked who their all-time hero in mathematics is, will cite Euler. And, of course, I would say that myself. His work just has enormous appeal, partly, as you say, because he doesn't get tangled up in all sorts of details in foundations.

Selberg: He operated at times rather recklessly, in matters of divergence, and things like that. But he had good instincts. So he didn't really go astray.

MP: He also had extraordinary taste in finding good questions.

Selberg: He was a remarkable talent and extremely prolific.

MP: There's another aspect of Euler that I like. When one reads his work, he talks about the problem, why he got interested in it, how he may have gone down the wrong path for a while, backed up, and started something else.

Selberg: That was a different time. These days if everybody started to do that, I think the editors of the journals would. . . .

MP: We can't afford it! But it does make more pleasurable reading.

Selberg: Oh, yes! That is true. It was much more human.

MP: Of course, it's also nice to see that even a giant in the field could start off with the wrong approach and be willing to tell you about it. Gauss was not.

Selberg: Gauss, of course, had a wrong philosophy. One has the impression, or at least I have the impression to some extent, that Gauss, when he wrote his *Disquisitiones*, was thinking of leaving something for number theory like Euclid's *Elements*, something that would stand for centuries. He does make a remark somewhere that he wouldn't expose the things that, as in a building, went into

Figure 14.3 The young Selberg family in the sixties—Ingrid, Atle, Lars, and Hedvig.

the construction. He was hiding his tracks, as other people said. This is a nonsense attitude, because something could only remain in the position of Euclid's *Elements* if mathematics stagnated for a long time. Of course, Euclid's *Elements* remained the same; nothing new was being done anywhere. This was thought to be the last word; nothing more could be done. It's obvious that as long as mathematics is thriving, the form that any specific paper or book or other form that mathematics takes is always advancing. The content remains.

Ramanujan

MP: *Let's go back to your school years. As a boy you had gone through a number of the formulas of Ramanujan before you even got to Oslo, to the university.*

Selberg: Well, yes, I started working on some of his formulas. I wrote a paper that I submitted that same fall that I entered the university. It was in German, the foreign language I knew best at that time. It was on some arithmetical identities, some connections between series and infinite products and continued fractions. It was published the next year. I had sent it to G. N. Watson—I think he was in Birmingham—who was the expert at the time on Ramanujan's *Nachlass*. It took him a long time to send it back, and I was rather impatient, I must say. Later, when I had to referee papers, I came to understand how that happens.

MP: *But when you got to Oslo, your dissertation ended up being on the Riemann zeta function.*

Selberg: Well, you see, I had gotten interested in mathematics through the work of Ramanujan and then later through some of Hecke's work in modular forms. So I was reading some number theory, specifically the Cambridge Tracts, like the one by A. E. Ingham. I never liked very fat books, so these Cambridge Tracts were the kind of books I could get something out of. I had started some attempts at work on the zeros on the critical line, using some ideas that were a bit different from the ones that had been used before. I had some results, but they didn't go so far as Hardy and Littlewood's. I was looking at their paper, particularly some end remarks they made—why they couldn't get more out of their method, and they gave some sort of explanation for that. I thought about that a bit, and then I realized that was really pure hogwash. That was not the reason that it wouldn't work. Basically it has to do with variations of arguments in small intervals or so. But it was not the argument, it was the amplitude, the absolute value, that fluctuated too much. So then I thought of introducing a damping factor that would dampen the fluctuations. I experimented first with very simple forms of the auxiliary factor that should dampen the fluctuations, but one that did not introduce any sign changes or anything. I gradually improved it, to see if I could get a better result, then I made a guess, and it gave me the final result. That gave me the dissertation.

No Dissertation Advisor

MP: *Who was your dissertation advisor, even nominally?*

Selberg: I didn't have any. We didn't have that system in Norway. You see, you wrote a dissertation entirely on your own. If you thought you had something good enough, you turned it in to the university, and a committee took a look at it. You didn't have to have any connection with the university to do that. You didn't even need to have a university degree in advance. Anyone could do it.

MP: *We certainly don't give degrees unless people pay tuition and lots of it!*

Selbeg: I know.

Figure 14.4 Selberg celebrating being named a Knight Commander of the Royal Order of St. Olav in 1987 by the King of Norway with his daughter Ingrid, wife Hedvig, and son Lars.

MP: *It's a different system entirely.*

Selberg: Besides the dissertation and the disputation in relation to that, you also have to give some lectures, one on a subject that you choose yourself and one on a subject given to you by the committee. I don't know whether that has been modified, but when I got my doctor's degree, that gave me the right to lecture at the university and the university had to provide me with the auditorium. On the other hand they didn't have to pay me! In Germany, there were lots of people with doctor's degrees who were not paid by the university but were paid by the students.

MP: *The privatdozents.*

Selberg: Yes. We didn't have them in Norway, but in theory, there could have been. I don't think anyone could have made a living at it.

MP: *There were a lot around Göttingen who were just barely making a living at it. It was a hard life.*

Selberg: Oh yes.

Prison Camp

MP: *Now, what did you do immediately after your doctorate?*

Selberg: I got my doctorate during the War, and I was a research fellow at the University at that time. But actually the University was closed in the fall of the next year, in the fall of 1943, not long after I had my disputation and got my degree. I was actually for a

while in prison because I happened to be in the University at the time the building was occupied by the Germans. But I was let out relatively early because I had friends in high places. There was a professor of actuarial mathematics with whom I had reasonably good relations before the War came, before the occupation. I had on some occasions actually helped him on some mathematical questions. He got involved in the Quisling party, and I think that it was his influence that got me out of this camp.

MP: *How long were you in the prison camp?*

Selberg: Not long, not more than one and a half months. Actually, about the same amount of time that I had been a prisoner of war in 1940. When I was released from the camp one of the conditions was that I shouldn't be in Oslo. So I went to live where my parents were at that time. I worked, and on some occasions I needed books from the University library, but I could not travel without police permission. I had to apply to the police for permission to go to Oslo to take out books from the library! I couldn't do it by writing since I didn't know exactly which volumes I needed. This was a bit cumbersome. The War did come to an end, however. I had been able to do some work in the meantime. I wrote some papers that I completed right after peace came in the spring of 1945, so I had a number of papers that I had finished and submitted in '45 and '46. In some ways, you see, even before the University was closed, the journals from outside had sort of dwindled down. First, things from abroad didn't come, certainly not from the countries that were at war with Germany, and that included eventually the United States. After a while even the German publications started to come at more and more irregular intervals. And then they stopped. So there was really nothing. Even *Zentralblatt* stopped coming. In a sense you were operating then without really knowing what was being done elsewhere, in more or less complete isolation. It was not a bad thing, actually, because you were left to your own devices. You didn't have that much distraction from periodicals that come in, and you feel that you have to look at them to see what is being done. Of course, you might worry that, my God, these things I'm doing might be somebody else's, and it's been done much better already. So I must say that after the War, when I heard that the Institute of Technology in Trondheim had gotten in things that hadn't gotten to Oslo yet, I took a trip up to Trondheim to look through them. I was pleasantly surprised. It seemed that nobody had really duplicated the work that I had been doing at the same time. Before the War I had had the misfortune of doing a number of things where it turned out that somebody else had done them just slightly before.

The Institute for Advanced Study

MP: *What prompted your invitation to the Institute in Princeton? Was it your dissertation, or was it the subsequent work that you did when you got out of the prison camp?*

Selberg: I think it was Carl Ludwig Siegel who probably helped—you see, he was a good friend of Harald Bohr. He had contacted Bohr, who had been on the committee for my dissertation. By the time my disputation came, Bohr couldn't be there because he was in Sweden. He had had to flee from Denmark. But he had been able to send his opinions, which were read by a Norwegian, Professor Thoralf Skolem. So I think it was Bohr who drew Siegel's attention to me. I must say, though, that I had met Siegel earlier when he came through Norway in the early spring in 1940, before he left for the United States from Bergen—on the last ship that went out before the German occupation. He was very lucky. It was certainly Siegel who wrote me to ask me to apply to the Institute.

MP: *It was also Bohr who wrote the citation in 1950 when you won the Fields Medal.*

Selberg: Well, I always thought that if Bohr had not been the head of that Committee, I probably wouldn't have gotten the Medal. I think these things matter. I don't know. For some reason, the first mathematical congress I took part in as a member—I went to some lectures at the mathematical congress in 1936 [in Oslo], but I was not a member—it was a congress in Helsinki (or Helsingfors as you said in those days), a Scandinavian congress, and I gave a talk as did my two brothers, twenty-minute talks, of course. Actually I was surprised I was received in such a friendly way by some of these older mathematicians, not only Harald Bohr but also by Torsten Carleman, who had much less

Figure 14.5 Selberg in his office in the Institute for Advanced Study (IAS) in 2005.

interest, I would think, in what I was doing. But he said some rather nice things about my work. I don't think I was that impressive at the time. I was just barely 21. No, I think had Bohr not been on the Committee, it might not have been me. Of course, had Bohr not been on the Committee, it might not have been Laurent Schwartz either, because I know Bohr was very impressed by the theory of distributions. I think that on the whole, except in some rare case where there is someone really spectacular—Serre maybe, who would have gotten it no matter who was on the Committee—in other cases it's a matter of taste. It's the same with the Nobel Prizes. It's partly a matter of chance. And in other cases it's clear.

MP: *Yes, absolutely clear. Pólya once gave me his list of the three nicest mathematicians he had ever known and Harald Bohr was on the list.*

Selberg: He was an extremely nice man, and it was such a tragedy that he died so early.

MP: *With all the applications of elliptic curves to problems in cryptography, it seems that suddenly number theory has become a branch of applied mathematics.*

Figure 14.6 Selberg at the IAS. He was called a "mathematician's mathematician" by Peter Sarnak of the IAS.

Selberg: It would have given great grief to Hardy!

MP: *That's right. Hardy would not have liked it at all. A colleague of mine, Paul Halmos, has said that applied mathematics is bad mathematics.*

Selberg: Personally I see nothing wrong with mathematics being applied, as long as it's good mathematics that is being applied.

MP: *Of course, that is Paul's point too. Did you know Hardy?*

Selberg: No. I never got to England until 1971.

MP: *And he died in '47.*

Selberg: I did meet Littlewood. He was rather old at that time, in his nineties.

MP: *Yes, Littlewood visited here in the late 1950s. Of course, seeing Littlewood was fine, but seeing Hardy would have been interesting because he was surely a colorful figure.*

Selberg: Of the two books that they wrote, Hardy's *A Mathematician's Apology* and Littlewood's *Mathematical Miscellany*, Hardy's book is by far the more interesting. There's no doubt about that. Now I must say of Littlewood's book, there are two editions, the original, and then there is one later where someone has added a lot of things. That was a mistake. It doesn't add to the value. Much of the new material is rather trite, and though Littlewood may have said some of these things, they did not originate with him. Some of them Littlewood had heard already from other people. Littlewood's book was interesting, but Hardy's was a great piece of literature.

MP: *I'm hesitant, though, to recommend Hardy's book strongly to my students. When he gets off into the part about his being so proud that he's never done anything that could be of use to humanity and so on, I wince a bit. I don't know how my students read that.*

Selberg: One must also wonder whether this might not have been written tongue in cheek, perhaps to irk some people. I think he liked to do that.

MP: *I think so. I detect in his writing an inclination to provoke. He must have had an interesting sense of humor. So many of the Hardy stories are quirky and interesting.*

Hardy Stories

Selberg: I heard a number of Hardy stories from Harald Bohr. Actually, once when I visited Bohr, the last time, in 1949, 1 was looking through a book that he had, where visitors had written something. Hardy had written about his experiences a good number of years earlier in the United States. He was in the habit of rating things.

MP: *His famous lists.*

Selberg: Of course, the Grand Canyon was rated very high. But there was another thing, in Princeton, that was rated high. It was a restaurant on Nassau Street that existed in those days. It was the first restaurant on Nassau Street that I ever ate a meal in, when I came to Princeton. It impressed me because the walls were all white tile, and it looked like a grand men's room that had been changed into a restaurant!

MP: *How attractive!*

Selberg: He thought that had to be high on a list; it was just what Cambridge needed. It must have appealed to him for some reason. Actually Bohr had a number of stories about Hardy, partly related to his atheism. It wasn't really atheism. His views were directed against God; they wouldn't have any meaning unless God exists.

MP: *There would be no adversary, no point to it!*

Selberg: Here's a story about Hardy: he was in Stockholm and went to visit the Cramérs. Harald Cramér's wife showed him into the living room where Hardy observed a vase where there was a single rose and a book of poems by Rabindranath Tagore. It was open to a page with a poem to a rose. Marta Cramér waited for him to take in this carefully arranged thing and say something. So Hardy said: "Rabindranath Tagore, the bore of bores." The perfect squelch.

MP: *Something of a disappointment for Mrs. Cramér!*

Selberg: Definitely.

MP: *Apropos his views of Tagore, I found a letter from Hardy among Pólya's papers that ranked poets. He gave one hundred points to Shakespeare, citing something in Macbeth, seventy-three to Milton, seventy-one to Shelley, thirty-nine to Tennyson, and twenty-seven to Browning. And he gave 0.02 to the American poet, Ella Wheeler Wilcox, who wrote "Laugh and the world laughs with you, weep and you weep alone." He may have thought of her in the same terms as Tagore.*

Selberg: My impression is, partly from that book he wrote, that he did have a good sense for poetry. And he had good taste.

MP: *He was a cultivated man.*

Selberg: Yes, and he had a good sense of language, much more so than most mathematicians.

Is Number Theory Too Hard Today?

MP: *I have heard people say that number theory is just too hard now. The big problems have been around so long and have been looked at by so many good people, what remains is just too hard.*

Selberg: One reason mathematics may seem very hard today is that, in a sense, too much is appearing all the time. It's hard to keep up with it. I think it's very essential to know what you should not read. It's almost more

Figure 14.7 Selberg with a camera-shy tortoise in the Galápagos in 1997.

important than knowing what you should read. It's very difficult to say whether the old problems are the hardest. Actually, the oldest problem in number theory is the problem of the existence of odd perfect numbers, which has not attracted much attention. No really notable mathematicians have spent very much time on that. It may be partly because they consider the problem an unnatural one. But a lot of problems are unnatural— the Goldbach problem, for example. That's certainly an unnatural problem. Why should you try to write even numbers as a sum of two primes? There is no earthly reason for it. Of course, that's what it has going for it, according to Hardy; it certainly couldn't have any applications! We are sometimes interested in problems just because they are hard! And sometimes if you succeed in doing one, you may have something that can be applied also to other things.

Still, there is something fascinating about number theory. I don't think people will stop doing it, but it is definitely very hard. People tend to concentrate on the hard problems. They attract attention. The Riemann hypothesis is one. But at least it is a problem with meaning, not like the Goldbach problem. It's a question you have to ask! There are a number of people working on it today, and some people think they are close to a solution. On the other hand, so have people thought several times in this century. I don't know—they may still think so well into the next.

MP: *Someone said once that number theory is that branch of mathematics everyone would work in if he or she were good enough.*

Figure 14.8 John Nash with Selberg on the occasion of his ninetieth birthday celebration.

Selberg: I don't think that is true. People have different kinds of talents and different types of intuition. There may be people who can do very fundamental work in topology and have a particular affinity for that and a special kind of intuition. They might not do very well in number theory or vice versa.

MP: *But even some of the silly problems we talked about—the Goldbach for one—have an appeal early on to people because people can understand them. It gets them interested in mathematics, but then they realize that they have to earn a living and they go on and do something else. I often hear that people got interested in mathematics through some little number theory problem.*

Selberg: Yes, that is true. I'm sure in past times many people may have gotten into number theory because of the Fermat equation. Gauss refused to work on it. But it did lead over the years to good

mathematics—Kummer's work, and, of course, the eventual solution, which came from completely different directions. It may be that way with the Riemann hypothesis as well; the solution may come through some connections that we have no idea about today.

MP: *I recall that at the 1996 Seattle conference on the Riemann hypothesis, where you were the keynote speaker, there were physicists and people from a wide range of mathematical fields who were working on the problem.*

Selberg: Sometimes a physicist's way of thinking can introduce some useful ideas.

Winning a Fields Medal

MP: *Let's move back to your Fields Medal. I recall Lars Ahlfors saying that when he won it in 1936, he scarcely knew what it was. People* *hadn't heard of it before. By 1950 people had heard of it, of course.*

Selberg: I think it had largely been forgotten. I knew of it at the 1936 Congress in Oslo. There I actually met Ahlfors briefly, because he was a friend of my older brother. But I had forgotten about the Medal. So when I was approached before the Harvard Congress and told that I would get one of the Medals, it was a surprise. It was a very happy occasion because I didn't expect anything. And for the same reason, there were many others who didn't expect it. The problem with any prizes or medals is that they make one or two people happy, but they make a much larger number unhappy each time. So in that sense any kind of prize overall increases human unhappiness!

MP: *Alas!*

Selberg: So it's questionable whether prizes are a good thing.

MP: *Later, you won the Wolf Prize, and that involves real money.*

Selberg: When I got the Fields Medal, there was some money with it. It was $1,000 Canadian. And at that time the Canadian dollar was roughly equal to the American dollar. Meanwhile, the value of the gold in the Medal has increased a lot. If it's made of the same purity today as it was then, and I would assume that it is, those have some monetary value. It's a very nice medal. But it's not the kind of thing you can hang around your neck.

MP: *You've done a number of things in mathematics that now carry your name,*

Figure 14.9 Atle and Betty Selberg crossing a Norwegian fjord.

contributing to the mathematical language: Selberg sieves, the Selberg trace formula, the Selberg zeta function. . . . Does that give you great satisfaction?

Selberg: Well, maybe, in a sense. Of course, as you know, in mathematics many things carry the wrong name. The Pell equation was not known to Pell. I recall that André Weil once said that if something is named for someone, in most cases that is not the person who discovered it. So we shouldn't pay too much attention to these things. But it is nice to know that some of my work has not been wasted. It seems to have been useful to other mathematicians. And it has certainly been useful to me. It has given me a fairly good living

MP: *It's not very nice for people to have favorite children, but one might have a favorite*

mathematical result. Do you have a favorite theorem of your own?

Selberg: I have not really thought about that. I don't think I really have a favorite result.

MP: *Nothing of which you are most proud?*

Selberg: No, I don't really think so.

The Riemann Hypothesis

MP: *Do you believe that the Riemann hypothesis is true?*

Selberg: Yes, I do.

MP: *At the 1974 Congress in Vancouver, I recall Hugh Montgomery's saying in his talk that he believes it on even-numbered days and does not believe it on odd-numbered days or the other way around. You believe it on all days.*

Selberg: Yes. At one time one might have had some doubts, but with the additional numerical evidence, I think there couldn't really be any doubt. And I would say also that it would really be a blot on mathematics if a zero turned up off the critical line. Contrary to what Leibniz thought, this is probably not the best of all possible worlds. But with regard to the Riemann hypothesis, God should have gotten at least one thing right.

MP: *So all the zeros are there.*

Selberg: They're all there.

The Institute— Anything Missing?

MP: *The Institute at Princeton, it seems to me—and I suspect to most academics—has to be viewed as heaven on earth. Have you ever wondered about what you might have missed not working in a regular university with students and the variety of activities that go on in a more usual setting?*

Selberg: There are two things about it. One thing is the missing students. The students are of various kinds. If you are in a university, one has many indifferent students, and one may have some good ones from time to time. And that is a great experience. I have missed something there. I have also missed something else: if you have to lecture regularly in a university you have to keep up with a number of things that you may tend to let lapse in your mathematical knowledge. Or you may refrain from learning some new things that you would have to learn if you were asked to give a course on it. So I may know less mathematics than I would have had I been a professor in a university. I would have had to keep up in areas that haven't really interested me that much. I've had contact with bright young people at the Institute. Some people come there at a quite early age. But they already have direction when they come to the Institute. You don't get the opportunity to shape their interests to the extent you can in a university. So I may have missed something there also. There is the fact too that if you are fifty years in the same place, you wonder whether it might not have been more valuable as a human experience if you had spent these fifty years at several different places, as most academics do. So I probably missed something, but I also gained something.

MP: *While at the Institute you've done quite a bit of traveling too. You spent some time at the Tata Institute in Bombay, I recall.*

Selberg: Well, many mathematicians do a lot of traveling. In some ways, I don't really like traveling that much. I don't like the process of getting from one place to another.

MP: *Being there is fine; it's the process of getting there that's not fun.*

Selberg: It's not fun. It might have been more interesting in earlier times, going in a leisurely way by ship or train. But one has to make do with the life one has had. I cannot opt now for anything else.

MP: *The other afternoon at the barbeque Brian [Conrey] and I were asked what our hobbies are. Both of us sat there and tried to think about our hobbies. I'm afraid that most of the things that I could think of—and I suspect the same may be true for Brian—are sort of attached to mathematics. I could think of collecting old mathematics books, and listening to music. What are your hobbies?*

Selberg: Well, I do have some. In my youth I was very interested in botany, and I collected a large herbarium. I don't collect plants anymore. But I'm still interested in them, and I probably know and can identify more species than most. I have collected seashells for quite a number of years.

Figure 14.10 Selberg on the Great Wall of China in 1998.

MP: *But all those spirals—that's mathematics.*

Selberg: Yes, but some don't involve spirals, the bivalves, for example. To a minor degree I've collected minerals.

MP: *But those are often polyhedra.*

Selberg: Yes, crystals are connected with group theory. That's true. In general I've been very interested in nature, in flora and fauna. That has always fascinated me. I've not been interested in physics, in spite of its being closer to mathematics.

MP: *I feel the same. I feel bad about it because I think I may be missing something, but. . . .*

Selberg: I had some interest in rational mechanics when I was at the university. But that's essentially just a branch of mathematics.

Mathematical Giants

MP: *We came close to this earlier on when we were talking about Euler, that he was a giant in mathematics. Now let me ask another question: who are the giants of twentieth-century mathematics?*

Selberg: I don't think I know enough mathematics to say. The ones who have meant the most to me have perhaps not been the giants. Hecke among German mathematicians of this century would be one; in a sense I appreciated his work more than Edmund Landau's, though I read a good bit of Landau. Hardy and Littlewood meant something; Ramanujan meant a lot. I think Ramanujan and Hecke influenced me most. I don't think I'm competent to judge who is the greatest mathematician of this century. Hermann Weyl was a very impressive figure. He was very different from Carl Ludwig Siegel. A lot of Siegel's work was the sort that when you saw it done, it still seemed sort of unbelievable afterwards. I think that Siegel's lectures may often have had a negative effect on the audience in the sense that they saw all of this tremendous power being applied, but afterwards it still seemed as impossible as before. On the other hand, most of Hermann Weyl's work was the kind where the problems seemed extremely hard, but once you have seen it done, it seemed rather easy and natural. These are probably the most important things in mathematics. In that sense, among mathematicians I have known, I would put Hermann Weyl at the top.

MP: *You wrote somewhere about the distinction between two types of mathematicians, the problem solvers and the theory builders—of course, there's a spectrum of people in between the extremes.*

Selberg: You can undoubtedly make that distinction. In the long run, probably the people who introduce new ideas and concepts that lead to new theories and new directions may be more important than those who solve some problem. Of course, it can depend on what goes into the solution. It's very hard to give a cut and dried answer to such things. In general, perhaps, it may be the things that lead to new theories, quite new ways of thinking, that are ultimately more important than solving specific problems.

MP: *In one sense Andrew Wiles just solved a problem, but what he brought together to do it was extraordinarily ingenious and may have far-reaching implications.*

Selberg: Oh, yes, that is true. There's no easy answer.

Fifteen

Jean Taylor

Donald J. Albers

Jean E. Taylor is a professor of mathematics at Rutgers University, vice-president of the American Mathematical Society, and a member of the board of directors of the American Association for the Advancement of Science. She finished first in her class at Mount Holyoke College. Taylor retains the questioning manner that distinguished her as a schoolgirl, as evidenced by her fondness for the bumper sticker "QUESTION AUTHORITY!" Over the past quarter-century she has studied minimal surfaces, the closed surfaces enclosing the least volume or those with a given space curve as boundary that have minimal surface area.

In the early days of her career as a mathematician Professor Taylor tested her theoretical models of minimal surfaces by dunking loops of wire into her kitchen sink of soapy water. Later she turned to high technology, especially computer graphics software, to model bubbles. Today, she has moved from soap bubbles to crystals, which conform to more complicated rules about minimal surfaces.

Taylor's daughter calls mathematics "the family business." Two of Taylor's three collaborators on a recent research project are in the family. With her husband Frederick J. Almgren of Princeton, her stepson Robert F. Almgren of the University of Chicago, and Andrew R. Roosen of the National Institute of Standards and Technology, Taylor is trying to model the growth of snowflakes and other crystals on a computer. Her daughter Karen Almgren, a senior at Princeton High School, gave a talk at the spring 1996 meeting of the New Jersey Section of the Mathematical Association of America: "Calculating the Capillary Forces Exerted by a Drop between Two Cylinders." It seems the family business is expanding.

Taylor loves the mountains. An accomplished climber, she was invited to join the Annapurna expedition. Which of her many achievements most pleases her? "The day I climbed both Cathedral Spires and Church Tower in the California Sierra."

"Scrambled Brains"

MP: *Where did you begin life?*

Taylor: I was born in San Mateo, California, and moved to Sacramento when I was

Figure 15.1 A very young Taylor enjoying the California sun.

three. I don't have childhood memories of any place but Sacramento.

MP: *What do your parents do?*

Taylor: My father's a lawyer, and my mother's a gym teacher.

MP: *Did either one have interests or abilities in the quantitative sciences?*

Taylor: I think I'm totally unique in that area. My sister is in languages, and my brother is also a lawyer. Both sides of my family are involved in education but not in mathematics or in science.

MP: *Did your parents play with you a lot? Lawyers and teachers can be busy.*

Taylor: I remember when I was in college, my father somehow latched onto this business that you can prove that you can't trisect an arbitrary angle. It seemed to him very odd that you could prove that you couldn't do something. So he would send me totally nonsensical drawings in his letters, saying "These are my latest researches on how to trisect an angle." It was a running joke between us for a great many years.

MP: *How about your mother? You probably spent more time with her.*

Taylor: Oh, yes. One of her great disappointments in life was that she could never teach me how to do a cartwheel. Here was this gym teacher with an uncoordinated kid! She's a very warm person, very good at relating to other people. She became a high school counselor after she stopped teaching.

Figure 15.2 Taylor—a happy school girl.

MP: *What kind of a kid were you?*

Taylor: I did like to read. I read an enormous amount—science fiction, Daniel Boone and pioneers, that genre. For a long time I felt as if I must have been born exactly in the wrong age. In the future I could have been the captain of a spaceship. And in the past, I could have been an adventurer as a pioneer. Sacramento and the early fifties just seemed entirely wrong; I was not in the right place.

MP: *I remember having similar desires at that age.*

Taylor: A few times I had accidents that knocked me out. When I was a kid I put a rocking chair up on the table on our patio and then turned over backwards and fell on the cement. When I was eleven I was in a bad automobile accident. I had a concussion with fluid coming out of my ears and was in the hospital for a week. My father said

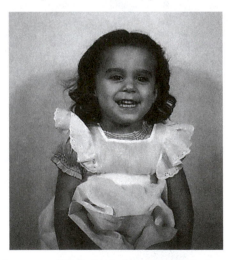

Figure 15.3 The look of a future leader. Taylor has worked on numerous committees and served a term as vice-president of the American Mathematical Society.

that each time my brains got scrambled they turned out a little bit better.

MP: *How about school itself?*

Taylor: In elementary school I got a C in handwriting and in self-control.

MP: *Self-control? How about academic subjects?*

Taylor: One time we had a math problem, multiply 26 times 40, and it had the 40 on top and the 26 on the bottom. So I flipped it over to do the multiplying. The principal gave me a very hard time about that: "You do the problem the way it's given to you. Put the 40 on top, and you must have the 26 on the bottom."

MP: *You wanted to take advantage of the zero.*

Taylor: Why, of course. It seemed the most obvious way to do it, to me. Another thing I remember very clearly is that I liked to do the puzzles, which they had at that time in the newspaper. Like, you have fifteen jars total, and there are twice as many jars on the second shelf as on the first shelf, and one less on the third.

I remember that algebra was fabulous because here was a means to make these problems really simple. I really appreciate this need—having a preestablished desire to solve that type of problem made me much more motivated to learn how. I think that was part of the issue of what my elementary school called self-control. I was pretty obstreperous in class!

Figure 15.4 "QUESTION AUTHORITY—That's one of my favorite bumper stickers," says Taylor. "It's one of the most important things to bring to mathematics."

In ninth-grade algebra I kept asking, "Well, why can't you do it this way?" At one point my teacher said, "I tell you what. I'll give you an exam in calculus that I took when I was in the Navy. If you do well, I'll give you an A. If you don't do well, I'll give you an F. You have two choices. You can accept it and take this exam or shut up."

I chickened out and shut up.

MP: I have noticed in some meetings that you certainly question very fundamental assumptions.

Taylor: One of my favorite bumper stickers is "Question Authority!"

MP: How about later? Those high school years can be ferocious.

Taylor: Something that made an impact on me in my midtwenties was meeting one of the women who had been really popular in high school. She had been class president, and here she was working as a teller in a bank. I had the feeling that her best years in life were probably over and that mine were obviously still in front of me.

Figure 15.5 Taylor continued to QUESTION AUTHORITY in high school and college. Although her academic grades were outstanding, she consistently received low citizenship grades. The Remarks section of her report reads: "Pops firecrackers in my apt. and hangs out with a bad crowd—probably is the ringleader."

High school can be so tough! Kids shouldn't worry so much about the popularity thing—it's not that big a deal. Still, you feel it so much when you're in high school. It's a whole lot better to have lots of good years later instead of three good years in high school.

Mountains and Mount Holyoke

MP: *So, in high school obstreperous Jean Taylor was pushing the outer envelope. You clearly did very well throughout your high school years, because you went from there to Mount Holyoke. How did you choose a women's college?*

Taylor: The only colleges I knew about in the East that accepted women at that time were the Seven Sisters colleges. You could apply to three for the price of one, which sounded like a good deal. Mount Holyoke had two lakes and a ski hill, so I put down Mount Holyoke as my first choice.

The other place where I thought I really wanted to go was Stanford. I got my acceptance to Stanford, and I was all ready to send it in when I got a telegram from Mount Holyoke saying that they had given me a Seven College Scholarship. I had never received a telegram in my life, and that really impressed me. They asked me to wire back, collect, and to keep it confidential!

MP: *You were really a hot ticket.*

Taylor: Yes. Also my father was changing jobs at that point. Up until then, my father had been with a bank, and he decided to go into private practice. The family income plummeted for a year or so. I got scholarship offers that evened out the cost, so even with two trips back and forth across the country to Mount Holyoke, it wasn't going to be any more expensive than going to Stanford or to Berkeley.

MP: *So you really got to choose where you wanted to go.*

Taylor: Entirely. I had never been east of the Rocky Mountains, so I decided to go to Mount Holyoke.

One thing that I have been impressed with since then is how well Mount Holyoke has done in graduating women who go on to get PhDs in science. I had no idea of this when I went there, but it was probably one of the best places I could possibly have gone. Just recently, I was looking at the absolute numbers of women who go on to get PhDs in the physical sciences and mathematics. The top four schools were Harvard, MIT, Cornell, and Berkeley. Tied for fifth were Rutgers and Mount Holyoke—a 1600-student women's college!

MP: *Can you recall any special reasons for choosing a women's college?*

Taylor: For me it was really odd because in high school my friends were almost all boys. There was a group of six boys and me who hung out together a lot. We ate lunch together, we all piled into a car to go off to football games together, and so on. I did not have any really close friends who were girls. Most of the girls in high school didn't seem to be interested in the things I was interested in.

MP: *Which was what? Football games?*

Taylor: We went to all the football games and most of the basketball games. Some of the guys were interested in science things, but none of them were top-notch students.

MP: *So you were definitely interested in science at that point?*

Taylor: Oh, yes. There were four of us who tended to win most of the academic awards; I was the only girl. So it was very weird at Mount Holyoke. I was surprised how hard it was not having boys as friends. Everything was set up in those horrible mixers! I remember showing up at Mount Holyoke and looking at a blotter on my desk, with a schedule of some of the Ivy League football games. I said to myself, "Gee, it would be nice to go," and then as the semester progressed I realized that you had to find a boy and have a date to be able to go to those games.

The thing that rescued me was the Outing Club. I got involved in that at the end of the first semester. Otherwise, I would have been ready to transfer and come back to California.

MP: *I noticed that when you were at Berkeley, you were in hiking clubs.*

Taylor: Yes. But now my husband, Fred Almgren, doesn't have as strong a passion for mountains as I do. We have a cabin in the Sierras, near Lake Tahoe, that my parents built with another family. It's near Mt. Ralston, which I have climbed many times.

MP: *So you really love mountains.*

Taylor: Oh yes! I remember at one point in graduate school my boyfriend asked if I could be doing anything, what would I choose? I said I'd like to be hiking in the mountains. And he said, "I would like to be doing exactly what I'm doing now, working on mathematics."

Good Chemistry

MP: *How did you end up as a chemistry major at Mount Holyoke?*

Taylor: The chemistry department was really exceptional. First of all, when I went there,

the chemistry department said: "Based on your mathematics, you should skip the first year of chemistry and study qualitative analysis." And I had only had ordinary chemistry in high school! Although I had completed a semester of calculus in high school, the math department said, "Well, we're not sure your math is good enough for us." So they put me back in math.

The chemistry department treated me more specially, which was good for my ego. And there was this dynamism that was just not there in mathematics. It was challenging and interesting. Anna Jane Harrison, who was the chair of the chemistry department, went on to be president of the American Chemical Society and the American Association for the Advancement of Science. She was a very forceful and intelligent woman. Another chemistry professor, with whom I did a couple of honors projects, was Lucy Pickett. She was really an exceptional person. So in chemistry I had all those exciting things to do with these really spectacular people. I liked it.

MP: *What did you like best about chemistry?*

Taylor: My high school chemistry teacher had suggested I think about chemistry as a major, and I said, "I don't like labs." And he said, "Well, in college you'll take enough labs that you'll learn how to do it, and you will enjoy it."

So there I was in qualitative analysis, where you have to work with samples. I extracted one thing after another, and while stirring things I managed to put the glass stirrer through the bottom of the test tube and got junk all over the top of the table.

All I could do was look at the colors as one substance reacted with another. There was no hope of going back through that whole procedure again.

Next I took organic chemistry. At one point, the professor was coming through, and I started to talk to him. I knocked over a graduated cylinder, lunged for it, and knocked the whole apparatus over. Glass flew everywhere.

I was not very competent in any of these lab courses, so my chemistry professors eventually told me, "Look, there is this thing called theoretical chemistry, where you can get out of the lab." I thought that was the key.

Finally, in graduate school my advisor told me, "A chemist who can't do experiments is like a man who deliberately cuts off one of his hands." According to him, I needed both the experiment and the theory. So it seemed to me that my time when I could get out of the laboratory was receding into the far future.

MP: *Well, if it's any comfort a few other well-known mathematicians started in chemistry and gave it up for essentially the same reasons: the laboratory was just an absolute disaster area.*

Taylor: It's too bad, because chemistry is full of fascinating ideas. This whole business about the shapes of molecules and how they react is great stuff.

Berkeley Years

MP: *Despite your laboratory problems at Mount Holyoke, you were elected to Phi Beta Kappa and graduated summa cum laude. Your chemistry grades must have been good enough to get you into graduate school at Berkeley, one of the best places in the country.*

Taylor: I didn't know what I wanted to do when I applied to graduate school. I applied to Harvard in biophysics, Berkeley in chemistry, and a bunch of different places, and I got in everywhere. But during my senior honors project on separation of the components of collagen, when I had to obtain the collagen by pulling out the tendons of frozen rats' tails, I decided that biophysics was a bit too messy for me. So I went to Berkeley in physical chemistry.

MP: *Back to the West Coast. Did you miss the West? Do you think that was a big factor in your decision?*

Taylor: No, I had also applied to Harvard in biophysics. I'd gotten more and more involved in mountain climbing and rock climbing when I was at Mount Holyoke. I went rock climbing a lot in the Shawangunks, and those mountains could have kept me happy.

MP: *So you went to Cal, where the laboratory scene was not an improvement over Holyoke. Did your mathematical interests begin to emerge then?*

Taylor: Yes. I guess the best way to explain it is I got involved with the hiking club as soon as I got there, and a lot of the members were mathematics graduate students. I found myself hanging out with them, and they would suggest interesting classes for me. So I began auditing some mathematics classes.

MP: *While hiking you would end up talking about mathematics?*

Taylor: The math grad students were talking about it, and I listened. Mathematicians everywhere talk about math while roaming.

MP: *But you must have had a fair amount of training to appreciate what they were talking about.*

Taylor: Graduate students are always interested in passing on what they've learned to other people. I had had two years of calculus, one year of abstract algebra as an undergraduate, and I'd been told by the chemists at Holyoke that group theory would be good for chemistry. At Berkeley I first audited a course in algebraic topology. Then I audited one in differential geometry by Chern, and it was fantastic. That did it.

The previous March I had passed my doctoral qualifying exams in chemistry, so I was just doing my thesis work. I found myself coming in every morning and not really wanting to get down to work. I'd read the newspaper for an hour, have coffee—it was obvious avoidance. I just wasn't enjoying chemistry.

Differential geometry, on the other hand, was full of beautiful ideas. So I talked to Chern about the possibility of switching to mathematics, and he encouraged me to do so. He also told me to take the final, even though I was only auditing the course. I did, and on the basis of that final he wrote me a

letter of recommendation for being accepted into mathematics at Berkeley.

MP: *Berkeley was a wild place in the 1960s.*

Taylor: That's for sure! When I went off to college, I think I was still a Republican like my father. During my time at Mount Holyoke I had been getting progressively liberalized, even radicalized. Then, at Berkeley, friends of mine were in teaching assistant strikes and stuff like that. I was involved a lot with antiwar demonstrations, even though I never worked harder in my life than I did that first semester in mathematics at Berkeley. It was terribly challenging.

MP: *Well, you didn't have a standard mathematics background for grad school.*

Taylor: I was behind. I read Walter Rudin's book and went back and reread linear algebra and I. Herstein's algebra and tried to fill in the other things. So I did a fair amount of reading and was working hard. But there I was, suddenly in math graduate courses.

I took functional analysis and tried to do a problem on the midterm—to prove that two norms were equivalent. I got a C, which was dreadful for me. When I talked to the teacher about it, I said, "Well, these are the words you're supposed to use, and I thought I was using those words." He looked at me and asked, "Do you really understand what it means for two norms to be equivalent?" I said, "No, I don't."

That was a real shock, to be in a hard, legitimate mathematics course. It was embarrassing! But somehow I managed to survive that chaotic quarter.

The next quarter, mathematics was more stable, but I was extremely unstable. My boyfriend John was writing me very nice letters from the IMPA [Instituto Nacional Matemática Pura e Aplicada] in Brazil. I was getting distraught at the political situation, and John wrote, "Why don't you come and join me in England? Get out of Berkeley and go to the University of Warwick and marry me." It was an appealing, romantic thing to do—so I did it. I dropped everything and was on an airplane to England within a week. At Berkeley I arranged to take rapid finals in various courses to get credit. I left for England just before the end of the second quarter.

MP: *To Warwick, which was essentially a new university.*

Taylor: Green and peaceful. Warwick and Princeton.

MP: *So there you were in Warwick continuing to study mathematics.*

Taylor: Yes—and hiking, naturally: Mont Blanc and other lovely peaks.

MP: *The Alps? Color me envious.*

Taylor: My mountaineering was pretty much limited to expeditions in August and skiing in the middle of winter. I also did some hiking and climbing in Wales.

Figure 15.6 On top of the world! Taylor is an accomplished mountaineer and was invited to participate in the Annapurna expedition. She is especially proud of the day that she climbed both Cathedral Spires and Church Tower in the Sierra Nevada mountains.

John was worried about getting his draft notice while we were there. He figured he needed to be more in touch with things, so he wanted to come back to the United States. So after a year and a half in Warwick, John went to work at the Institute for Advanced Study. I applied to Princeton, the graduate school closest to the Institute.

MP: *Had you met Fred Almgren by that point?*

Taylor: The first time we met was in 1970 at the International Congress of Mathematicians in Nice, just when John and I were going to Princeton. His talk on geometric measure theory was wonderful! It was just the sort of thing I was interested in.

I spoke to him afterward, and he described a problem: There are three half-circles at 120 degrees, and the surface of least area spanning them is three half-disks. Then if you move any one of the curves an arbitrarily small amount, it wasn't even known if the curve down the middle stays a finite length. It was very appealing to me.

MP: *So when you got to Princeton, you had a nice problem to work on.*

Taylor: Yes. By that time I had already had four years of graduate school: two in chemistry and two-thirds of a year in math at Berkeley, and one year at Warwick. I had passed two different sets of PhD qualifying exams—in chemistry at Berkeley and in math at Warwick. I'd also studied for the one at Berkeley and had intended to take it, except I left for England. So I figured I'd been a student long enough!

At Princeton, I put my blinders on and finished my thesis in two years. John had finished his two-year term at the Institute, so then both of us were looking for jobs. At the very beginning, MIT offered John a position and me an instructorship. Stanford was going to split one job between us. And Santa Cruz was offering something like one and a half jobs for the two of us. All the negotiations seemed so complicated.

MP: *This was 1972? The job market was really falling to pieces.*

Taylor: Yes. But we had those two independent job offers from MIT. I could appreciate all the efforts to accommodate spouses, but it certainly feels as if you're being dealt with in a more straightforward manner when you get an independent offer.

Unfortunately, during that summer weaknesses in our marriage started showing up, and in the fall we split up. Later, we separated for good.

MP: *When you were a graduate student in mathematics, there were considerably fewer women in the field than today.*

Taylor: There were no women faculty at Princeton, that's for sure. Though among the graduate students that year there were actually a lot of us: Marjorie Stein, Marsha Simon, and two others.

MP: *Were other people not especially supportive of women in a graduate program?*

Taylor: I felt quite intimidated by the other graduate students. The common room conversation and things like that were not encouraging. But I just didn't pay a lot of attention to it. I'm really good at putting the blinders on. Furthermore, I was there only two years.

Putting the Blinders On

MP: *Tell me about putting the blinders on.*

Taylor: I can really put them on. For example, I can be reading and be so focused I simply don't hear people around me talking to me.

MP: *Well, those who have looked into the work habits of some famous scientists say that some had profound powers of concentration. Newton apparently could work on a problem single mindedly, quite unaware of the most basic needs. He'd even forget to eat. It sounds as if you're very fortunate in having that power of concentration. It would be a great advantage in working on a tough problem.*

Taylor: My family doesn't necessarily think it's fortunate! I find it very distressing if I'm trying to focus on something and somebody wants to come up and talk to me. It takes me a long time to climb back out of what I'm focused on and then a long time to get back into it. So I absolutely hate to be disturbed when I am really concentrating.

MP: *So how do you work?*

Taylor: I like to find chunks of time so I can really sink into a problem.

Visual Images

MP: *Do you deal primarily in visual images?*

Taylor: Oh yes! I have to. To really understand something I need pictures of it.

That's one of the reasons abstract algebra never appealed to me. Polynomials never had any good pictures to go with them. With some of these approaches to crystal growth, like the viscosity solution method, I can "see" it.

When people just write formulas on the board and prove things by manipulating the symbols, that does not appeal to me. It was very hard for me to get the feeling of what polynomials are all about. Only if I try to understand it in my own way can it move beyond just the symbols to the point where I really understand why it works.

MP: *Understanding in your "own way." When do you feel you really understand something?*

Taylor: When I know why it works. Not just the logical chain, but what's behind it.

MP: *Your work is tied closely to the physical world, so the "why it works" can take on an extra dimension for you. You can literally see some of this.*

Taylor: Yes, but there's a physical why and also a logical why. One of the things I really liked in my initial contact with differential geometry was having concepts like torsion, which provide a language to describe physical properties of curves. I thought it was wonderful.

If you have the language of curves without pictures of curves or mental images of curves, that won't do you any good. But to have the curves without the language is also no good. It's the interaction that makes the difference.

Experimental Math

MP: *In a recent article, "The Death of Proof,"* you are referred to as an "experimental mathematician." Do you accept that appellation?*

Taylor: Yes. I like what Steve Krantz said, that David Hoffman and I do experiments, but we also prove things. If you just do experiments without trying to give proofs, then you're missing some of the heart of mathematics. I absolutely agree with that. I do some experiments, maybe drawing little pictures on paper or manipulating images on the computer. I work from examples in order to understand.

All the current ideas about how you should teach certainly resonate with the way that I learned things, which is understanding the example as a motivation for ideas. Good examples can really embody the heart of the theorem. I do experiments, for sure. But I also try to prove that what I see is what you have to get.

MP: *As I look around the mathematical community, I don't see many mathematicians who are able to play with wire frames and soap solutions, or with crystals. There's something really pleasant, I think, about being able to look at a frame in a soap solution and see your theoretical results.*

Taylor: Yes, it is pleasant. On the other hand, I'm not sure that playing with soap films gave me any of the insights for my thesis or the work I did after that. Still, having seen soap bubbles for so long, maybe it was ingrained. I knew what the surfaces looked like. I didn't have to study them carefully. I don't recall ever having tested something as an aid to conjecturing.

MP: *But it started with bubbles, and you later hooked up with metallurgists. Where are you going to go next?*

Taylor: Well, there are still a lot of other interesting things out there in terms of the shapes that crystals make when they grow or shrink. About five years ago I was really wondering where to go next, and then I found myself in the middle of all this stuff involving crystal growth. I did not know much of anything about it. At the moment I'm co-authoring a couple of papers, and I'm way behind in doing my share. Right now, it's more a matter of how I'm going to get through doing all the things I'm supposed to do in the next few months!

What I've been looking at recently is surface diffusion on these new materials. I'm doing this with curves, which are much easier than surfaces.

MP: *It doesn't sound as if you're going to run out of problems too soon.*

Taylor: Not in the next year or two.

*John Horgan, "The Death of Proof," *Scientific American* 269, no. 4 (October 1993): 92–103.

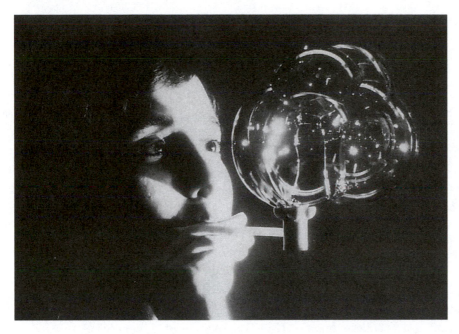

Figure 15.7 Looks like fun! Mathematician Taylor loves bubbles, for they provide beautiful examples of minimal surfaces.

Good Problems

MP: *What's a good problem for you?*

Taylor: One of the most important things for me in choosing a problem is whether I can make a unique contribution. If it's something somebody else can do, and I'm just trying to race to do it faster, that's never had any particular appeal for me. I want something that's not too likely to get done unless I do it.

But it also has to have some geometric aspect. I would find it very difficult to do something without any mental pictures. I certainly like the connection with physical reality. That's very appealing.

One thing I enjoyed that surprised me was some historical research. Someone had pointed out that the things I was talking about were like "zonohedra." I looked up H.S.M. Coxeter's definition of this strange

term and found that his definition had an extra condition in it that I didn't want to be there. I checked to see where it came from and traced it to the Russian crystallographer Federov. So there I was, reading a German translation of a Russian article, trying to understand it with my somewhat incompetent German. And the extra condition wasn't there!

So I traced back in Coxeter's articles and found where early on he had overlooked something and said that this thing had that as a consequence. Then he started taking the consequence as part of the definition. From that experience I can understand why people enjoy historical research.

When I checked with Marjorie Senechal, she went back and read the original Russian and said "You're absolutely right. That's what Federov said." So the zonohedron originally meant what I wanted it to, but it had acquired this other meaning, and I couldn't have it back again.

Lots of Balls in the Air

MP: *Besides being a woman in mathematics with a very good research record, you have been the vice-president of the AMS, you serve on a zillion committees and advisory boards, and your involvement is not likely to go down. In fact, the pressures are likely to increase. Why do you do it all? Does it feel good?*

Taylor: I've never done just one thing. In high school I was involved in a huge number of activities. When I got to college, out of the way of my family, I stopped going to church and I stopped playing the piano, after taking piano lessons for ten years. I just totally stopped all of the things that I had been doing. But I put other things in their places. On most weekends I was off rock climbing or mountain climbing or skiing, and I also was politically active. Ideas about how you can shape politics were very appealing to me.

What kicked in my political interest most solidly was the assassination of Jack Kennedy. That really had a big impact on me.

MP: *Do you remember what you were doing when you heard about Kennedy?*

Taylor: Of course. I was in the physics lab.

I've been interested in politics a lot for a long time. Between my junior and senior years in college I was invited to be an intern in Senator Cranston's office, although I chose a chemistry summer program at Mount Holyoke instead. I went to the inauguration of Johnson and Humphrey in 1964, and in the inaugural parade I had a sign from a Humphrey rally I'd organized at Mount Holyoke. Humphrey saw it and waved at me.

MP: *All of those activities sound slightly daunting.*

Taylor: I don't think I've ever been single-minded in doing anything. Plus it's more interesting to have lots of balls in the air. Though sometimes they all seem to come raining down, and it seems a bit much.

MP: *You've got a lot of them in the air now.*

Taylor: Yes. I've also got two graduate students working with me now, and that's wonderful. One of them just proved her first theorem.

I have said no twice to being a candidate for department chair. And I said no to some other things. I don't want to be an editor of a journal because I know that is something I would not be good at. If you send me papers, they will pile up in my office. So there are certain things that I do manage to say "no" to.

MP: *At age fifty, you seem to be picking up steam. For a number of mathematicians at the half-century mark, it's often going the other way.*

Taylor: In some sense, that is a concern—whether my energy is going to stay at a high

Figure 15.8 Taylor, technically retired but still doing mathematics.

level. I always thought in high school and college that one reason I did particularly well is that I worked hard. I did try to manage my time reasonably well. It sometimes seems to me now that I get tired more easily. Maybe I'm not getting enough physical exercise. But lying around for a while can sometimes seem awfully nice!

MP: *Well, you seem to be accelerating.*

Taylor: Maybe. I like being involved in a lot of things. It's part of what keeps me energized. I probably would prove more things if I weren't involved in other activities, but I find my life more interesting if I am. If I were trying to maximize the numbers of papers I publish or the depths of the papers I publish, I would lead a different life than I do. This underscores something very different about Fred and me, because he is much more focused on doing his work than I am. It is not nearly as important to him to have other things to be involved in.

Gender Issues

MP: *What do you think women bring to mathematics that men don't?*

Taylor: One important thing anyone can bring to mathematics is the "right" to question authority. Just trying to look at things differently is actually where you get your insight. Some of the people who have made the big strides did not just take the received wisdom and accept the same modes of thought. Instead they asked,

"What if you look at it this way?" If you come into a situation thinking about things in a way different from anybody else, that can be a plus.

To the extent that women are socialized differently, they can bring a different aspect and identify what questions are important to them. People talk about the male idea of dominating the environment. Women don't think about science that way and don't approach it that way.

Most of the women I know of who are good at mathematics are also good at a whole lot of other things. The men who rise up in mathematics are often more focused on one thing, or maybe they can somehow afford to have one interest and be much more into doing mathematics. Take, for example, talks. I listen to lots of professional talks, and women give better talks. Women on the committees I have been on have worked much harder than the men— on the average. Maybe this is an aspect of our mothers making us more rounded than men.

MP: *Or perhaps women feel the pressure to perform well because they're "representing" women in mathematics.*

Taylor: It does seem to me that the women I know in mathematics have a greater richness than the men I know. I'm not talking about all men or all women, of course.

MP: *Two years ago we did a survey of the entire MAA membership, and about 10 percent responded. It was very apparent that on*

average, by several different criteria, women seemed to focus more on teaching and students than did men.

Taylor: I think that's true. It has always seemed to me that there is something exceptional about the general performance that women put forth in mathematics. A lot of weeding out goes on, and perhaps it is only women with relatively strong personalities who manage to stick with mathematics and research, and they can handle lots of things. If you haven't got high energy levels, you are much less likely to make it. And therefore you're interested in education, in many things.

MP: *Do you think male faculty members, when considering a woman candidate, still engage in some of the old rationales? Saying, "Well, we know that she's just going to go off and have babies and not really be an active mathematician." Is gender still a big factor?*

Taylor: I think it is. On the other hand, this is a blinders issue. I don't think it does any good for younger people to stay focused on that. It just takes up energy that is better spent elsewhere.

MP: *For a time consciousness was high, but there's a danger that it will fall back.*

Taylor: Well, have you seen Beth Ruskai's statistics on participation of women at AMS sessions? Notice the number of sessions with zero women—even one that was organized by women had zero women. Thirteen organized by men had zero women. I posted the statistics on my office door and somebody

came along and wrote "Good job!" Somebody else had written up in the corner, "I'd rather be shopping."

Role Modeling

MP: *Your name is all over the place these days, and, like it or not, you have become a role model for a lot of younger women who may be aspiring to careers in the mathematical sciences. You may not have done that by design, but it's a fact.*

Taylor: When I think of some of my women professors in college, intentionally or not they had effects on me and served as models.

MP: *I know that you are brought into contact with women students who are thinking about careers. What do you tell them? What's your*

Figure 15.9 Sometimes mathematicians work on other problems, too. In this case, daughter Karen's hair is the problem. Jean is assisted by her stepdaughter Ann.

advice to women who are thinking about serious work in the mathematical sciences?

Taylor: The key is finding out what really turns them on. You listen and try to see what makes their eyes light up. And then you encourage them to do what they want to do.

Sometimes I advise them that having blinders is not a bad idea, but simultaneously I try to convince them that you don't have to have blinders.

I strongly emphasize that they should do what they want to do. Today, students also are concerned about finding jobs. My hope is that they also become aware of how mathematics ties to other subjects and that the current talk about interdisciplinary work will not be all just talk.

MP: I have the impression that you are not too bullish on mathematics education these days.

Taylor: I'm enthusiastic about trying to teach. I wish the students were as enthusiastic in return! There are some students in almost every class for whom I seem to make a major difference. And that makes a big difference to me. I have students from ten years ago who still write back on occasion.

MP: So you really like teaching.

Taylor: Yes. But I could strangle those students who refuse to learn.

MP: What is your explanation for that kind of behavior?

Taylor: I really don't know. For some students there is just nothing that I do that makes a difference.

MP: We haven't talked much about your children, yet.

Taylor: The importance of my kids, including my stepchildren, to my life has been huge. A big part of the pleasure in my life is having children. Maybe women are not supposed to talk about that, but it's a real good thing for me.

I'm very proud of all my children. My stepdaughter Ann and I hiked the John Muir trail together, and she is an utterly delightful hiking companion. It can be quite an experience to spend three weeks with one person.

MP: I understand that your teenage daughter took complex analysis at Princeton last semester, while still in high school.

Taylor: Karen is a smart kid! Absolute straight As in high school. At age sixteen, she was taking abstract algebra at Princeton. I can hardly believe that my own daughter likes abstract algebra! She jokes about mathematics being the family business. I'm absolutely astonished at this child Fred and I produced.

Unexpected Results

MP: You've done many things—and not just in the mathematical sciences. What do you feel best about?

Taylor: Well, I remember one spectacular rock-climbing day in Yosemite with Greg Schaefer. In one day we climbed Church

Tower and both Cathedral Spires—the three big climbs. That was a definite high point!

Then there have been various times of getting wise, figuring out some mathematical things. Sometimes they were not entirely positive experiences; what I expected did not turn out to be the case.

MP: *You mean you got a particular result and it was different from what you had guessed? Is that a little discomforting?*

Taylor: Well, it shows that my assessment of the situation was missing something. It says: "*Look, kid, you were wrong.*" On the other hand, it's also very exciting!

Postscript: Fred Almgren, Taylor's husband, died in February of 1997. After retiring from Rutgers University, she moved to New York City, where she is a visiting scholar at the Courant Institute of Mathematical Sciences of New York University.

Sixteen

Philippe Tondeur

Donald J. Albers

Philippe Tondeur is a research mathematician and a consultant for mathematics, science, and technology. His current interests include mathematics research and engineering; innovation policy; institutional governance; and leadership development.

In 2002 he retired as Director of the Division of Mathematical Sciences (DMS) at the National Science Foundation (NSF). Previously he had served as head of the department of mathematics at the University of Illinois in Urbana-Champaign (UIUC).

He earned an Engineering degree in Zürich, and a PhD in Mathematics from the University of Zürich. Subsequently he served as a research fellow and lecturer at the University of Paris, Harvard University, the University of California at Berkeley, and as an associate professor at Wesleyan University before joining the UIUC faculty in 1968, where he became a full professor in 1970.

He has published over one hundred articles and monographs, mainly on differential geometry, in particular the geometry of foliations and geometric applications of partial differential equations. His bibliography lists nine books. He has been a Visiting Professor at the universities in Buenos Aires, Auckland (New Zealand), Heidelberg, Rome, Santiago de Compostela, Leuven (Belgium), as well as at the Eidgenössische Technische Hochschule in Zürich, the École Polytechnique in Paris, the Max Planck Institute for Mathematics in Bonn, Keio University in Tokyo, Tohoku University in Sendai, and Hokkaido University in Sapporo. He has given approximately two hundred invited lectures at various institutions around the world.

He has served as Managing Editor of the *Illinois Journal of Mathematics*. Professor Tondeur has been a recipient of the UIUC Award for Study in a Second Discipline (Physics), a UIUC Award for Excellence in Undergraduate Teaching, a Public Service Award from the Society for Industrial and Applied Mathematics (SIAM), and the 2008 SIAM Prize for Distinguished Service to the Profession.

Professor Tondeur recently chaired the Board of Governors of the Institute for Mathematics and Its Applications at the University of Minnesota. He served on the national Advisory Council for the Statistical and Applied Mathematical Sciences Institute at the Research Triangle Park in Raleigh, North

Carolina. He has also served as a member of the National Committee on Mathematics of the U.S. National Research Council. He is a member of the International Scientific Advisory Board of the Canadian Mathematics of Information Technology and Complex Systems (MITACS) Centre of Excellence and a Trustee of the Instituto Madrileño de Estudios Avanzadas-MATH (IMDEA-MATH) in Madrid. He is a member of the Committee on Science Policy of the Society for Industrial and Applied Mathematics, as well as of the Science Policy Committee and committees of the Mathematical Association of America.

In the interview that follows, we learn about the unconventional path that he took to becoming a mathematician. We also explore his decision to take on academic leadership. According to Tondeur, "Leadership is about accepting responsibility and acting on it. It's not about getting a position; it's about using a position for impact." He may surprise some mathematicians with his claim that "research is a wonderful preparation for science administration and leadership, especially the failures."

MP: *You were born and educated in Zürich. Did you grow up speaking French and German?*

Tondeur: French at home, German in school.

MP: *So you are trilingual?*

Tondeur: Well, my grandmother spoke Italian to me, and later I learned Spanish. Fluency in English came later. My first book in English was Halmos's *Finite-Dimensional*

Figure 16.1 Tondeur (left) at age fifteen with his older brother and nephew in 1947 in Zürich.

Vector Spaces. I deciphered it with the help of a dictionary. I still admire Halmos.

MP: *What did your parents do?*

Tondeur: My mother came from a family of teachers, and my father worked at a bank. I had an undirected childhood in terms of intellectual pursuits, also known as total liberty. I came from a modest economic stratum of Swiss society, and there were no attractive models to emulate. My own formation came through reading. I read voraciously as a child, and now that I am retired I again read voraciously.

MP: *What did you read?*

Tondeur: Everything, I had no guidance. My first extended view of the world and life was through books.

"I Found Geometry, and It Was Wonderful"

MP: *When did science become an interest?*

Tondeur: Very late. I basically stumbled into it. I started with a technical career and served an apprenticeship in a factory. I recognized that behind all this technology there is theory, which attracted me, so I did study engineering (in night school) and earned a degree in mechanical engineering. I won first prize in my graduating class, which encouraged me to continue on my journey. While progressing toward that degree, I recognized that there was a much wider world out there. I gravitated toward the world of science. My driving impetus was always more meta-thinking, more abstraction, and mathematics was the natural convergence point.

My initial journey within mathematics was of the same nature: toward the more abstract. Within mathematics I also meandered around in complicated ways, but finally I found geometry, and it was and remains wonderful. While delighting in almost all forms of mathematical thought, I settled on differential geometry, for it was just about the right mix of holding onto the world in a robust way but also in a wonderfully abstract way. It was just the perfect territory for the way my mind works. The larger framework is topology, and in later years, I got fascinated by geometric aspects of partial differential equations. In fact, that's basically where I have done all my research over forty years, while being interested in almost all forms of mathematics.

MP: *You said that most of your study had been undirected. Were there teachers who stand out in any special way, who encouraged you and said, "You're a good student, you should do more of this?"*

Tondeur: There were a few key voices that said, "You are good at math," and this was an important encouragement—"Oh, OK, I'm good at math." First there was Viktor Krakowski as math teacher in my preparation for university admission (I ranked second nationally in that year's Abitur exam for students seeking university admission outside the public school system). At the University of Zürich there was Rolf Nevanlinna, who said at one point, "You should have an academic career." And I was totally astounded. He said, "You are a natural for it." So I became his assistant. I was thinking of perhaps going into teaching, but I was not really contemplating an academic career up to that point. I searched for quite a while. Where could I find some activity that would be commensurate with my dreams? I had huge literary interests and continue to have them, as a reader but not as a writer. Mathematics was wonderful—a mix of contemplation and action. I'm not technically brilliant, so I had to find a subject that allowed my imagination to flow freely, and differential geometry was just right. There is a natural affinity. I suppose it is like a musician who discovers an instrument which speaks to him, and he embraces it for life.

MP: *In talking with mathematicians and asking them how they think about problems, we hear many of them say, "Well, I try to develop some kind of geometrical picture."*

Tondeur: Definitely. Everything has a geometrical aspect for me. It took me a long time to find out that's what I really loved and was good at. You find a pursuit in concordance with your natural disposition, and you are a very lucky person. I am a geometer.

MP: *Nevanlinna was a famous person to be working with. What was it like?*

Tondeur: Huge! Dominant. I mean, not so much in technical terms, but here was this Prince of Mathematics and this glorious subject that prominent people were treating as an important life pursuit. It was his personality, his eminence—the weight of this intellectual persona. I thought, "Oh, so it's a good subject; I could spend my life doing it." Prior to Nevanlinna, I had met other excellent teachers, but with them it wasn't so seductive. And historical figures were too remote.

A Peripatetic Mathematician

MP: *So you finished your PhD in 1961.*

Tondeur: Actually, I had just about nine months for the whole PhD studies after my diploma. Nevanlinna said that I should become his assistant, and I started working on a problem and solved it. That was it.

MP: *That's fast.*

Tondeur: Then I went to Paris on a Swiss Science Foundation fellowship, and thus my postdoctoral wanderings began.

MP: *You were a bit of a peripatetic mathematician there for a few years.*

Tondeur: Yes. I had no plan. I was simply searching for a place that would be intellectually meaningful. By then it had to lead to an academic career, and where would that be? I had no idea. America was beckoning, but France was interesting, so I went to Paris, and then I got an invitation from Harvard. In my ignorance, I had no concept of Harvard, but it was an offer (from Raoul Bott), and so I came. I asked for a visa and came to the United States and never went back. Everything was accidental.

MP: *There's an important factor here, I think, that made it easier for you in some sense to come to a different country, namely your command of languages.*

Tondeur: Yes, it was an essential factor.

MP: *Someone who's been educated in the United States would not necessarily think about going to France or Germany.*

Tondeur: I have taught in four languages—in Spanish in Argentina, in English in the United States (and the United Kingdom, New Zealand, as well as other countries), in French in Paris, and in German in Switzerland and Germany—but language never was an issue. The dominant issue for me was: Where could I act optimally? I didn't think about it strategically but, rather, instinctively. You have to do things commensurate with your talents and limitations and fulfill . . . some dream. So, I tried to assess where the impact would be positive. You do things,

Figure 16.2 The dashing Tondeur at age twenty in Zürich.

achieving which will lead to further goals. In that sense, I was not really ambitious, just evaluating opportunities as they arose. Prestige did not enter my thinking—yet I was an egotistical young person, I wanted a good job, but I wanted a job where I could do good at the same time. I didn't really aim for the highest imaginable position; it also had to be compatible with what I felt I could deliver.

Actually, in all my wanderings I always had the sense that I should complete what

Figure 16.3. Iurie Caraus with Tondeur at the International Congress of Mathematicians, 2006, in Madrid.

I had set out to do. I would first fulfill the set task, and I did lots of things which were sometimes not optimal careerwise, but I did them out of a sense of duty which comes from my Swiss background.

Figure 16.4 Tondeur at the University of Bonn, 1967.

MP: *You are describing a strong work ethic?*

Tondeur: When I accept responsibility for a task, I try to complete it. It doesn't matter if it's profitable or not. If I signed up, I try to do it until completion. I've had this discipline, which is ingrained and which comes from the Swiss educational system. It was beaten into me.

MP: *So what eventually attracted you to Illinois?*

Tondeur: An offer that looked attractive—a tenured offer.

MP: *I noticed your first appointment was as an associate professor.*

Tondeur: While I was at Harvard, Shiing-Shen Chern invited me to Berkeley. In my first course assignment, I replaced Charles Loewner who fell ill and taught my first differential geometry course. Then Wesleyan University offered me a tenured position as a start. Illinois looked great, and I had visited prior to their offer. It was a big university with many eminent people. It turned out to be a fabulous place.

MP: *So once you landed in Illinois, you didn't leave except to come to the NSF.*

Tondeur: Yes. I had several offers over the years, but I didn't see any compelling reason to change. Moreover, Illinois was very liberal in its leave policy (thank you, Paul Bateman), and I served in many visiting positions all over the world. By a rough count I gave about two hundred lectures on my research interests of the moment in many countries.

Figure 16.5 Professor Emeritus Tondeur at age 70.

There is fantastic freedom in a huge place where people can pursue many interests, so that was the charm. As chair (much later), I could similarly accommodate people wanting to do different things. In a small department, you are more constrained.

MP: *So you started at Illinois in 1968, and it was in 1996 that you became the department chair. I know that you were a distinguished teacher as well as a researcher. When did your interest in teaching emerge? Was it always there?*

Tondeur: Always. I spent a great amount of effort on teaching, or learning through teaching. I would try to understand where my students came from, what their level of preparation was, and then I would try to do the best possible teaching job within these constraints. If they didn't have the background they were supposed to have, I would teach it. I like teaching; it is fun, and I liked it because I paid attention to the students' needs. I was not a very standard teacher. I got lots out of the ordinary assignments. That was the wonderful thing about Illinois. If you do it, OK, we'll find a way to accommodate you. I had fantastic freedom. I repeated very few courses. Most courses I taught just once. It was an incredible education. I have wandered through vast stretches of mathematics; yet mathematics is so immense that I still feel that I grasp only a small part of it.

Avoid Boredom

MP: *Do you think you're the sort of person who gets bored easily?*

Tondeur: Never bored. I mean, I avoid things that bore me.

MP: *Yes, but how about the fact that you're constantly moving on to teach new courses, although you say you're also doing that to learn new material.*

Tondeur: Yes, I seek out new things. I try not to be in situations which are repetitive.

MP: *Let's go back to Switzerland for a minute. I've known a few other mathematicians from*

Switzerland and in particular from Zürich, and they have been very effective teachers. They regarded it as a good activity. Do you think that Protestant work ethic is part of the explanation for that? Let me be more direct: Are mathematicians from Switzerland likely to be more attuned to teaching than mathematicians from other countries?

Tondeur: I don't really think so. When I studied there, there was a tradition of teaching being important, but this is not unique to Switzerland. This country's (the United States) attitude toward teaching is changing. We are in a corrective phase, after some excesses, and teaching is becoming again more important, as it should be. For eight centuries, universities have been about teaching. What's the highest degree at the university? It's a doctorate. What is a doctorate examination? You have to teach your research. Your results are accepted after you present your research to a committee. So teaching of your research is the alpha and omega of academic life. It's been true since the first university was established, and it's true in the American research university too, even if it has sometimes been forgotten.

Contact Hours—Wrong Measure for Teaching

MP: *There are various explanations for the apparent decline or de-emphasis on teaching in research institutions in the United States. I remember talking one day with Al Tucker, who was the chair at Princeton for twenty years. He was there from 1933 to 1970. If you look at the collection of people who were on the faculty then, it was a world-class group.*

Tondeur: Stellar.

MP: *Of course, he cared very much about teaching, and you can see that passed on to his two mathematician sons, Alan and Tom. I asked him what Princeton was like in those days in terms of the educational aspect of their program. He said that teaching was very important, and there was none of this business of people always seeking ways of reducing their teaching time to do research. He said, you taught—this I found absolutely mind-boggling—in those early days, you taught fifteen contact hours, and you did your research.*

VIGRE—Reminder of Fundamentals

Tondeur: Princeton continues to have a high appreciation of teaching. I don't know about every single case, but it has had extraordinary success in inculcating the value of teaching in its faculty and students. I think Princeton has continued to have a high teaching ethic and has fantastic role models, for example, Ingrid Daubechies, Charles Fefferman, Peter Sarnak, and Elias Stein, among many others. They are great teachers who love teaching with a passion. Princeton is still a model. The number of contact hours [hours in the classroom] is a poor parameter of teaching. We probably have to get away from that. It's not a meaningful measure. It's a natural thing when you seek a job that you prefer a lower number of contact hours,

but that's not the right measure because your teaching takes many forms. I think for the people of quality we are looking for in a research university, it is people who should have a sense of responsibility for teaching. Princeton faculty do this. They generate other such people. I had and have such colleagues in Illinois who were definitely shaped by Princeton, and they are fantastic examples of how teaching is crucial. Today's university mantra is the integration of research and education. Teaching is not separate. We teach through research. That's the model we are pursuing. Activities like NSF's VIGRE program (Vertical Integration of Research and Education in Mathematics) are very much about reminding the math faculty of the United States that teaching is a crucial part of their mission.

MP: *You speak with great passion about this integration. I certainly had some professors who said that research and teaching go hand-in-hand and that it's very hard to do one without the other, but you articulate it with passion. Please expand on this a bit, because I know you worked very hard on the VIGRE program.*

Tondeur: Actually, Don Lewis invented this program, and when I came to NSF, it was already in place. All I had to do was to build the Division of Mathematical Sciences (DMS) budget to be able to fund it (as well as a new program that I initiated, the Focused Research Group program, and many other things). The background to VIGRE was the assessment of the Odom Report. Basically, the mathematics discipline in the United

States has a problem. We don't recruit enough people. Fulfilling this need is part of the national agenda and the mathematics community's responsibility. Fantastic new mathematics is being created, but if you don't have the people, the future is at risk. I took the existing VIGRE program and worked on it. The NSF spent a lot of money on it and tried to promote and expand it later under successor solicitations with different acronyms but the same general aim.

Fundamentally, it is about the mathematics research faculty of the United States taking responsibility for the pipeline in the mathematical sciences. They should show the intellectual landscape to their freshmen because then they will carry that interest and awareness into their sophomore year. The reality is that we lose so many of those students who enrolled as freshmen, young men and women who came with the idea that they know math and might want a career in it. But after two years we have this "Valley of Death," this huge attrition, and I think that it's our responsibility to keep students interested. We get them for two years, and we have to succeed better in keeping them. That's our responsibility: we can't blame it on drugs and alcohol and Facebook.

We have it in our power to influence them. Why do they disappear? Why do they not go on? VIGRE is giving you resources to achieve a better result. That's one level. Where do the graduates come from? They come from undergraduates who decided to do more math and go to graduate school. And then when you get them as graduate students, you have another immense opportunity to treat them well and do fantastic things with them. And then they get their degree, and there is yet another level—what will they do now? They have to prove themselves and get a career going. That's a postdoc, right? And VIGRE attacks these three points in different order, while the focus is on the graduate students.

I hear a lot of comments to the effect that it's an immense task in addition to everything else we do. But I think it's central to what we do; it's not *in addition*. The successful VIGRE programs are those where a significant number of the faculty have taken responsibility to act, and to mentor, and to show the glory of the math discipline. That happens at all levels, the upper undergraduate level, and then graduate and postdoc levels, and VIGRE has had significant impact. In some sense VIGRE is really a reminder of the fundamental responsibilities of research faculty. The NSF is responsible for the discipline, and under the leadership of Don Lewis, its DMS decided to start this program. We used research money to do this, not waiting for some miracle education money to show up, and we have devoted lots of energy and money to it and its successor programs. It promotes cultural change. Some people complain that it's directive. I prefer to say it's a reminder of fundamentals, reinforced by grant money. It's essential for the discipline.

MP: *So you'd like to see a reawakening, a reestablishment of what may have been a tradition of the past?*

Tondeur: Yes, I would see it as a corrective. Research and higher education are integrated activities, and VIGRE is a reminder of our responsibility in this respect. Now we cannot, as research faculty, do everything in mathematics education. But we have lots of undergraduates who show up in our schools, 16 million of them, with 2 million enrolled in undergraduate mathematics courses. How come we don't retain more of these students? I think many of these young people walk away from potentially fantastic careers because they do not find out in their short time what it is we do as research mathematicians. Many people do very well with it, but we need more of them. My perspective is, from a national point of view, that we don't do nearly enough.

MP: *So VIGRE, it appears, is making a difference.*

Tondeur: It is. But there are many DMS activities aiming toward the same goal. VIGRE and its current successor programs support graduates, postdocs, and research experiences for undergraduates. DMS supports them and individual investigators in many different forms. These programs have multifaceted aspects. Many successful research agendas have support money embedded toward these very same agendas: VIGRE is just one very visible component of DMS's investments.

MP: *This is addressing the pipeline issue in a very fundamental way. Occasionally I hear others talk about the obligation that mathematics departments have to people who are not necessarily going to be finishing graduate degrees. We saw the Rochester experience* where the graduate mathematics program was almost eliminated. That was very painful. A few months ago I was talking to a fairly well-known mathematician who somehow got involved with some mathematics for business students. Of course most math departments view business schools as beneath them, or composed of people whom they don't necessarily regard as academic. Over the last fifteen years or so the profile of business students has risen dramatically, for the simple reason that they can go out and make a lot of money. This individual talked about a conference that was set up for business school deans, and the topic was mathematics for business students. On most campuses the relationships between business departments and mathematics departments are not very good. If a member of the mathematics faculty has been assigned a course in business mathematics, other department members will often say, "Oh you poor guy, that's terrible." And the person teaching it will often come back and say, "This is terrible. These students aren't with it, they're not thinking very much about the actual content." What can we do to be more relevant in some sense to what it is they're doing, because we know that business uses very sophisticated mathematics these days? Deborah Hughes Hallett was one of the people at the conference. She actually was asked to send out invitations to quite a large number of deans. She found that the interest level was very high, and she said it was very uncomfortable to turn people away. She had to send out "so sorrys" to over thirty business school deans.*

Tondeur: Amazing.

MP: *They came, and there weren't many mathematicians there.*

Tondeur: A missed opportunity.

MP: *Well, they wanted the business people to talk with the deans in particular. Deb said that more than one of these deans spoke to her in social settings, saying "Why have you invited us? Why do you want to talk to us? On our campuses we can't even get the math faculty to talk to us." So the upshot of the meeting was some rather good communication. That's just an example, but I think it bears on the pipeline question, if we are not connecting with the so-called client disciplines.*

Integration of Research and Education and Its Impact on the Economy

Tondeur: I think of the pipeline issue in the broadest sense. I don't think of the pipeline as only leading to the production of research mathematicians. I think of the pipeline as creating people who are comfortable with mathematical thinking and technology in all fields. That's the broad pipeline issue. My dean at Illinois wanted these people to be taught by mathematicians, not by journeymen instructors, because, I think, it's again the issue of integration or research and education, ultimately impacting our economy.

Mathematics is a key technology in our knowledge economy. Many new techniques require an enormous amount of mathematical sophistication. It relates to our pipeline issue. In the university environment, the math department has to be involved in this, and from my experience at Illinois, we have been. We are very interested in this in a purely pragmatic fashion of faculty positions, tuition income, and general financial support, but also intellectually because that's within our mission. It's getting better in this country to recognize this responsibility of the math profession, to take a hand in all this. If we don't, we kill our future, and we kill our resources.

MP: *There's another powerful force to deal with that seems to be growing, and that is the fraction of our faculties that are part-time.*

Tondeur: That's a bad development, but I have no solution to propose. I understand how it's economically driven, and I think it's a negative development. The equivalent in high school education in this country is people teaching math and science without disciplinary qualification, and I cannot think of this as a good development. What I could do as the chair was to just refuse to use part-time faculty. I give credit to my predecessors in the chair's office at Illinois who just wouldn't allow it, and we didn't do it.

MP: *So there was a tradition.*

Tondeur: Yes, a tradition that you have to maintain and continue to argue for—we didn't want part-time faculty in the classroom.

MP: *Illinois should be congratulated.*

Becoming Chair at Illinois

Tondeur: The dean gets credit here because he has to put up the resources. And the chairs have to fight for these resources. Deans also have a vision of intellectual quality—it's not all about money. In summary, I have nothing positive to say about the part-time trend.

MP: At Illinois, you eventually became the chair. How did that come to pass?

Tondeur: Somewhat reluctantly. Research faculty do not really like to take on this responsibility, and so chairs are recruited by colleagues and the dean. In 1996 I was one of the key people involved in the discussion about who would take on this job, who would be condemned to take it, and I was reluctantly drafted by my dean with the recommendation of the faculty as the least evil choice. It wasn't something I had planned to do. The dean said he would like me to think about this job, so I said yes, if that's what you ask me to do, and here are the issues. I was willing to negotiate about resources, and it dragged on for a long time. Finally he committed to fully fund what is now the postdoctoral J. L. Doob Research Assistant Professor program as a condition for my accepting the position of chair.

MP: With a quarter of a million dollar price tag?

Tondeur: In three permanent annual increments. Over the three years of my tenure as chair the dean put up a multiple of this in new resources for the department in order to support all that was needed.

MP: You said that you were brought to the position somewhat reluctantly and that you would be considered the "least evil choice."

"It's All About Leadership"

Tondeur: It is of course more nuanced than that. Before I accepted the position, I had communicated essential departmental needs to the dean.

MP: You did that, and I think that's a big plus. But there is a problem, and it's not just with mathematics departments by a long shot, because in some sense you're elevating the importance of leadership in a mathematics department. That's very hard to attack because the general perception of most mathematics faculty is that it's the last thing they want to do. They're not likely to be rewarded for being leaders, they may not even be respected as much anymore because their colleagues are far more interested in research. Although it's common for research mathematicians to be producing somewhat fewer papers as they get into their fifties and sixties, by that age many attitudes toward departmental leadership are fixed.

Tondeur : Yes. It's all about leadership.

MP: How do you attack this?

Tondeur: Leadership is about accepting responsibility and acting on it. It's not about getting a position; it's about using a position

Figure 16.6 "Pushing an agenda in a recalcitrant environment"—Tondeur's playful characterization of some leadership experiences.

for impact. I think of academic leadership as being in charge of a field of action, similar to being in charge of a research agenda. It means you are going to work toward realizing a specific agenda, which has to be embedded in the general strivings toward the ideal of the university. The chair is there to improve a specific department. So, it's an action program. I had no interest in the title. The main issue is the acceptance of responsibility. If you are not willing to do that, then you shouldn't become the chair.

For me the process of being recruited consisted of working myself to the point where I would want to do it. You have to think out a realistic action program which will further the goals of the enterprise. It takes a while: it took me two months to figure out before I was ready to accept. It is similar to working on research problems. You have a conjecture, and therefore you have to work on this and improve that, and so on, and you need specific tools, and finally you achieve some of it, not all of it. I know where I want to go, and I also have the experience from research that you don't necessarily achieve what you set out to do. What you find is actually different, but you are quite happy in the sense you have a goal, and by doing steps toward that goal you find out that actually it is different from what you thought, so you have to be flexible. I think research is a wonderful preparation for science administration and leadership, especially the failures. You fail mostly in research. You try thousands of things: most don't work, but a few do, and those successes are your achievements and define your life's work. Research is a tremendous preparation for leadership.

MP: *Most leaders don't talk much about failing.*

Tondeur: Yes, but it's essential to know how to deal with failure. Often you don't succeed, so leadership is also about dealing with failure; it's about restructuring the issues so that you can succeed. Our published mathematical research record is only about the successes, and so I think dynamic leadership is very similar to research. Now, what's different in a leadership position is that you have to engage other people. That's a new element; it's not you alone. That's a big difference.

MP: *You're good at it.*

Tondeur: I turned out to be good for the two specific leadership positions I was entrusted with in my career.

MP: *What's your secret?*

Tondeur: I find people interesting. I can appreciate a wide variety of people with diverse talents. So, for me, the question is how can people who have these skills or particularities or even quirks be mobilized to use their special talents for the purpose of the enterprise? It's like directing an orchestra. You have these different instruments, and I'm delighted that they are so different. Oboists are famous for eccentricity. It's wonderful to try to make great music out of the whole ensemble, including oboists (musicians will understand). I don't like slackers, but basically if you concentrate on those who are willing to work and if there are enough of them, then you'll succeed.

MP: *You've obviously had success at Illinois as a chair.*

Tondeur: It is what brought me to the attention of people searching for a new director at NSF's Division of Mathematical Sciences.

MP: *So NSF knocked on the door.*

Tondeur: Yes, Margaret Wright asked me to think about this job. Again, I was reluctant.

MP: *How long did it take you to make a decision?*

Tondeur: At the time I was asked, I had achieved an initial huge success at Illinois, and in the process I had developed an enlarged sense of responsibility to advance the profession. I thought that if I could climb this mountain, maybe I could climb a higher mountain too.

"I Brought My Checkbook"

MP: *Could we expand on the huge success?*

Tondeur: Transforming the state of mind of the department. It sounds a bit arrogant, perhaps, but it was a despondent department, and it became an excited department. That's a major change. It regained its faith in its destiny. The postdoctoral program brought new blood to the department, and then I took down a wall. We needed a seminar room, but we couldn't find one. I said, two of these small rooms will be the seminar room. Rip out the separating wall. In a hundred-year-old historic building (majestic Altgeld Hall), you don't rip out walls, so there is the

symbolic significance. I needed funding to do it. I brought my checkbook to the critical meeting and said if nobody else pays, I'm going to write the necessary check. I finally didn't pay any of it; the dean paid for it.

MP: *You felt very strongly about it.*

Tondeur: I was willing to write the check. It's only because I was willing to that I didn't have to. I expressed priorities: I put money in the terrific research library collection; we got this new seminar room; we got postdocs; and we recruited well. And my successors creatively built on and enlarged that success.

MP: *You've mentioned some of the things you did to revive and to inject a new spirit.*

Tondeur: Money is important.

MP: *Can you tick off a few other things?*

Tondeur: The recruiting of new colleagues. It was constantly on my mind. We made a huge effort. My first recruitment year was a fabulous success.

Figure 16.7 The Brothers Tondeur (Michel, Edmond, and Philippe) in Zürich in 2007.

The National Science Foundation

MP: *So you accomplished a great deal in a very short period of time, actually. You had a plan.*

Tondeur: Yes. Faculty. What's next? Students. Postdocs. What else is important? The library, a seminar room, online teaching, the merit workshops. So I had a list, and we made big headway in a very short time. So when the call came from NSF, I already had a second year of big successes, and I was thinking much about the mathematical sciences in a broader way. My experience in the mathematical sciences has continually reinforced the perception of the university as part of a worldwide network for scholars and students. Illinois is just one node, so in the case of the call from NSF, I was being asked to do this for the United States piece of the global math network. It didn't take me long to think about it. I sent them my dossier, and then nothing happened, and I went my way without thinking further about it. It turned out to be a long process. I said, "I want to finish three years at Illinois. I'm not going to leave before that." I thought, once I had given three years of leadership to my department and had been successful, it seemed like an interesting challenge to try to broaden this activity. There also was an urgency of time because I was getting older, and if I wanted to do something outrageously ambitious, I had to do it now.

MP: *You don't act very old.*

Tondeur: Not the spirit, but the body is creaking.

MP: *So the NSF opportunity represented a chance to do more with the network? Scale it up?*

Tondeur: Scale it up. Mathematics is a global science. I have this life experience of the universality of the mathematical sciences. If you look at my bibliography, you will find 17 different research collaborators, and they come from around the world, so for me it's global science. Working at Illinois is working for the progress of the mathematical sciences. This NSF job was an opportunity to do the chair's job on a national scale. The worldwide scale would be another, but there is no platform for that. The United States has been the dominant place for science, and it impacts the world in a big way. Everybody's watching what's happening in the United States, so prioritizing mathematics and statistics in the United States has a huge impact on the support for the mathematical sciences throughout the world.

MP: *Did you know Don Lewis before coming to NSF?*

Tondeur: No, but he was a well-known figure. Once I was drawn into NSF I talked often with him.

MP: *How about Rita Colwell, the director of NSF? Did you know her?*

Tondeur: Not at the time, but she did appoint me as director of DMS. Once the search was finished, recommendations for director were submitted, she interviewed me, and thus my relationship with her started.

MP: *Her inclinations toward mathematics seemed to be quite positive.*

Tondeur: Yes.

MP: *Was that any kind of factor in your thinking?*

Tondeur: No, because you have to interact with whoever is in charge. You have to be pragmatic. But Rita Colwell turned out to be a great supporter of the mathematical sciences. A critical factor for our success was the existence of the Odom report, issued in March 1998, with the cumbersome title "Report of the Senior Assessment Panel of the International Assessment of the U.S. Mathematical Sciences." The report committee was chaired by William E. Odom, Lt. General, USA, Retired, former Head of the National Security Agency. (He died in 2008.) This report was the result of his and Don Lewis's initiative and leadership, and I was lucky to come in waving this report.

MP: *Well, in looking back, what in your NSF experience, as the director of DMS, has given you particular pleasure?*

Tondeur: Working with talented people toward a huge goal. And the huge goal is to allow the mathematical sciences to play their proper role in science and society. For science, mathematics is a driving force, and it is critical for the future of society, globally, not just for the United States. And education is a part of it. The biggest satisfaction is to advance the recognition of mathematics as a fundamental force and the heart of civilization.

(The interview continues in 2005.)

MP: *What are your thoughts on current NSF funding?*

Tondeur: My readings, observations, and travels over the three years since leaving the DMS have convinced me that we are not even remotely doing enough to live up to our current leadership role in science and technology. We were thrust into this post–World War II leadership role by visionaries advancing the public investments in our future with boldness, and with the scientific community's sense of common purpose, this advocacy was very successful. The results have been truly magnificent. They have massively contributed to our prosperity, health, and security.

Basic research is the driver of innovation. Knowledge creation translates into economic growth. Much of this innovation has spurred the current worldwide dynamics in science, engineering, and technology. But this global spread has turned into an economic race in which we are not running at our full capacity; in any case we are not running sufficiently fast to live up to our leadership responsibility.

"We're Not Investing Enough in Research and in People"

MP: *What are your thoughts about U.S. science leadership?*

Tondeur: There is evidence that the U.S. leadership in science and technology is eroding.

Our economic strength is based on past superior innovative capabilities. How is this going to look in the future, when the United States is currently seventeenth in the world in the proportion of college-age population earning science and engineering degrees? We are not investing enough in basic research, and we are not investing enough in people. The country needs to act more boldly on its science and technology agenda. There is no way to balance the federal budget by squeezing investments in our scientific future. Our investments in basic research as well as mathematics and science education need to increase at a high rate. The NSF budget request for fiscal year (FY) 2006 is close to $3 billion below the authorized target in the "NSF doubling bill," passed December 2002.

MP: What about the mathematical sciences?

Tondeur: Funding research in the mathematical sciences has fared relatively well at NSF, doubling from 1998 to 2004. But even NSF's support of the mathematical sciences as a priority area is faltering in the budget request for FY 2006. It does contain language about the mathematical sciences being a priority area, but not a single additional dollar is budgeted for DMS above the previous year's level. For a funding agency, this is talking but not walking.

MP: What about the longer term?

Tondeur: Funding for the mathematical sciences will keep pace with expanding needs only if the trend in federal funding of basic research takes a dramatic turn for the better (this is 2005). What is needed? I believe that

a threefold investment in NSF's portfolio is a target to aim at. An incredible amount of first-class research is left unfunded. Many mathematics and science education innovations are not tested out in pilot programs for future widespread implementation in our faltering public schools. This ultimately means an underinvestment in the development of our scientific workforce at the very time when a soon-to-retire scientific workforce is going to have to be replaced. At the same time it is a fantastic opportunity to advocate for a massive national effort in mathematics and science education and to involve an increasingly diverse population in this national effort.

MP: What about the role of professional associations? There are the Mathematical Association of America, the American Mathematical Society, the Society for Industrial and Applied Mathematics, the National Council of Teachers of Mathematics, the American Mathematical Association of Two-Year Colleges, the American Statistical Association, and many more. What do you think they might be doing that they're not doing to improve the place of mathematics in this world culture in order to convince this larger public that mathematics is really important?

Coordination—the Weakest Part

Tondeur: I think the coordination is the weakest part. I think it's a cacophony. All the players pursue very well-meaning, very sound, very good-looking small agendas, but there is no coordination to make it more successful. What would make it more successful is if there were more coordination, and that seems to be the weakest part of the effort. All these voices are there, but they don't produce a sound which registers on a political level or in the general consciousness of the public. People don't hear what we are discussing here. The general public, the Congress, which reflects the general public, only vaguely hears this noise, and then they forget it and get on to other more urgent-looking business. We haven't been successful in getting this basic message across, or not successful enough. I think there is an increasing sense, an increasing awareness on the part of many math career people, frustrated by more and more evidence of lessening mathematical university courses, quotas, and so forth, that having a national agenda for increasing awareness of mathematics is something important. But our general society is incredibly slow in registering this concept and its importance. Reading is something you cannot live without. But mathematical thinking is something you will not be able to live without in the future either. It's of huge significance, and we have not been successful in convincing the general public of this fact. This is not a criticism of what individual groups do; they do it admirably well with an enormous sense of responsibility, but there is no integrated effort to do that. It is in the minds of the individual players, but you don't hear the big voice.

MP: *Over the past several years there has been increased interest expressed by different societies about getting the ear of Congress and getting on their radar screen. I've also heard*

Figure 16.8 Tondeur, a delegate of the International Mathematical Union, meeting with the U.S. delegation to the International Congress of Mathematicians, 2006, in Santiago de Compostela.

some say that the only thing that's really going to sell to Congress is research—but it's going to have to be directly tied to science because Congress doesn't understand mathematical research. Then others have said that if you're thinking about Congress there is one part of the whole mathematical enterprise that most of them think they know something about, and that's education. But that's pooh-poohed by many, particularly those from research institutions, so I think what you're saying about some coordination here could go a long way. Maybe the way to get to Congress is by coming through an education portal.

Tondeur: I wouldn't underestimate Congress even if the process is slow. There are things it responds to better, but as with most initiatives, it requires an intense education process. The math priority area, as conceived at NSF, was talking about three things: namely, the development of the discipline, which is a research agenda; the connections with science, which are the interactivity and interdisciplinary science-driving force; and education. All these need to be integrated. The most important thing we do as researchers about education happens in the classroom, seminars, working groups, and through our mentoring. Our players are active at this level, and that's what we think and do most about. It has a huge impact on the whole enterprise. It's not "If you do this, then you don't have to do the other, or only do research"—these activities are all integrated. Congress is responsible for the entire national agenda, so you have to impress on its members the importance of the special mathematical sciences agenda. It's natural that for Congress the education agenda is the most obvious aspect. But it's also about the workforce, jobs, the economy, tax revenues. It is all integrated. You can't have good mathematics education without the Andrew Wiles and Grigory Perelmans of the world. We provide support for the whole spectrum of activities. All the professional societies you mentioned provide contributions at some point of this spectrum. Better coordination would be that each advocate understands that he or she is part of this immense enterprise, which is not centrally directed. It is itself a complex system. Each participant should be imbued with a sense of common purpose of the mathematical world and mathematical thinking.

MP: *What preliminary steps would you take?*

Tondeur: We all have employment in a specific place. We act locally wherever we are positioned. I happened to be the director of mathematical research, so I did and continue to do it in my retirement through research advocacy. We all have roles to play. If you are a school principal, then you have to implement it in some way, mainly through recruiting the proper talent. I don't see it as one first step; I see it as this group of people pursuing the multiplicity of agendas, cognizant of the fact that it's part of a national purpose, or even a worldwide civilizational purpose.

MP: *But you still have to develop that broader appreciation to achieve coordination.*

Tondeur: You have to, and some people are very good at it. The Joint Policy Board for Mathematics has cultivated lots of people through the National Mathematics Awareness Month and its awards for outstanding communication. These are important contributions. It's a big agenda, and you push it at every opportunity, and you have to awaken the sense of responsibility and also a sense of common purpose and progress in this enterprise. I'm impressed by the work of all these organizations and the dedication of their members. What is not there is the resonance of an orchestra. Take biology; nobody doubts that biology is big stuff. It changes our lives, prolongs life, and one big message is: it doesn't happen without mathematics. You won't have those drugs that make you live longer without math. Take cell phones: what is going on with telecommunications? Now the message is encryption. A research group proved in recent years that primality testing can be done in polynomial time. This result potentially puts our whole system of secure communication at risk because of that theorem of mathematics. That fact was established by three Indian mathematicians. Aren't we number one in the world? Not necessarily. Great ideas show up anywhere. Collectively we haven't succeeded in having the coordinated voices in place, so our Washington presentation of these agendas has not been able to make the case in a compelling fashion. The sound of the orchestra is still very faint.

MP: We do gather quite a few mathematics societies together often in this building (MAA headquarters), in an umbrella group called the Conference Board of the Mathematical Sciences (CBMS). There is, of course an executive officer. But I'm not sure that the role of the executive officer was ever created with the idea of being a conductor. A conductor has a lot of power. It would take a remarkably diplomatic executive officer to lead this orchestra of societies. But at least there is a structure that gives you a chance.

Tondeur: You have to use the platforms you have: CBMS may be a platform that can be used in an effective way. I think there are very few members of Congress who would recognize CBMS. They might associate it with a news channel or a pharmaceutical company.

MP: It's much worse than that. I'm afraid the majority of mathematicians don't know what CBMS is.

Tondeur: What is the challenge of public math advocacy? By way of analogy: How long did it take for reading to become recognized as an essential skill? Not too long ago, maybe 150 years ago, most people didn't read; 100 years later, almost everybody was reading. Newspapers made it happen. All of a sudden the literacy level went up with the advent of newspapers. It is technology; you have news, and you print it and hand it out, and then reading becomes indispensable. We have to do this with mathematics. It is similarly becoming an indispensable part of everybody's life.

(The interview continues in November 2007.)

A Wake-up Call for NSF

MP: What has happened since we last spoke in 2005?

Tondeur: Almost three years ago I summarized my concerns as expressed to you in 2005 in an opinion piece entitled "A Wake-up Call for NSF," published in the June/July 2005 *Notices of the AMS*. This got many discussions going on these issues with opinion leaders in Washington and throughout the United States and practically identical issues with science and government representatives from several countries. This has led to continued travel all over the globe. Usually the discussion was framed in the larger context of the science and technology agenda for a specific nation or region. This has been an interesting experience about the political process, both here and abroad.

Let's return to changes in the United States: a political storm was triggered by the National Academy of Sciences report titled "Rising Above the Gathering Storm: Energizing and Employing America for a

Figure 16.9 Tondeur addressing the International Congress of Mathematicians, 2006, in Madrid.

Brighter Future" (aka RAGS Report) written by a distinguished committee and brilliantly co-chaired by Norman P. Augustine and P. Roy Vagelos. This report was requested in a bipartisan fashion by Senators Lamar Alexander and Jeff Bingaman, drafted in ten weeks, and came out late November 2005. Its gist was adopted by the White House and Congress. It has led to dramatic changes in the budget outlook, in particular for the mathematical and physical sciences. All this was beginning to become budget reality for FY 2008 (which in the United States began in October 2007, but the Omnibus budget bill in December 2007 in fact threw out all these expectations). In the summer of 2007 Congress passed, and the President signed a three-year authorization bill on math and science with the wondrous title "The America Creating Opportunities to Meaningfully Promote Excellence in Technology, Education, and Science Act"—or, "The America Competes Act" (a marvel of congressional acronym dexterity). While this is an authorization bill, the real money is in appropriations bills. Nevertheless, the outlook for federal support for these agendas over the next few years is improved, but as the saying goes, Congress has not only to talk the talk but walk the walk. The mathematics and statistics community has to seize these opportunities and turn them into a paradigm change for the role of the mathematical sciences in science and society.

International Consulting

MP: *Could you expand a bit on your work with other countries? When we talked recently, you spoke about their different structures supporting science and operating procedures and how you approached those challenges.*

Tondeur: Instead of expounding on general principles, let's take a concrete example. During the last year, I have been involved with the issue of the interdisciplinarity of the mathematical sciences as practiced in Japan. Mathematical research in Japan is practiced at an extremely high level, yet it has been observed by many that the interactions of mathematics with the sciences and with industry are much weaker than in other Organization for Economic Cooperation and Development (OECD) countries. A forthcoming OECD 2008 report spells this out explicitly, and I consulted with staffers preparing this report, in Germany, Japan, and the United States. I was asked to help on this issue with the development of specific proposals to change the situation.

I view these activities as a gardener looks at his opportunities and responsibilities. Through the accidents of my career, I have

Figure 16.10 Tondeur with Ari Paptev at the International Congress of Mathematicians in Madrid, 2006.

been thrust into a position where I am asked to attend and help in the nurturing of the garden of mathematics, one of mankind's most glorious creations. How could I resist, having been given the incredible privilege to spend most of my life in this garden? It is a privilege and an honor to be asked, and I do my best to help promising projects succeed (mainly in the United States, Canada, Spain, and Japan, but also in other European countries and Australia). I have served on the Boards of Mathematical Sciences Institutes in the United States, Canada, Japan, and Spain.

"The Mathematical Sciences Are Humanity's Heritage"

MP: *Are there lessons to be learned from or issues in working with other countries?*

Tondeur: Each country has its own (and sometimes no) science policy machinery.

My work with other countries in support of the mathematical sciences has raised an eyebrow or two, especially among my friends from the field of national security. Here are my thoughts about this. The sketchy list of names like Thales, Euclid (origin of pure mathematics), Archimedes (embracing both pure mathematics and interdisciplinary fields), Galileo (promulgating mathematics as the language of nature), Newton, Leibniz, Euler, Laplace, Lagrange, Fermat, Galois, Gauss, Riemann, Poincaré, Cartan, Einstein, Hilbert, Weyl, von Neumann, Wiener, Chern, and an ever-larger number of contemporary luminaries says it all: the mathematical sciences are humanity's heritage, and the development of it continues at a fantastic rate. The United States has been a stellar steward of this patrimony since fascism forced the emigration of leading scientists from Europe, and the U.S. generosity in providing them opportunities to flourish made the United States the leading supporter of this most brilliant of mankind's

Figure 16.11 The Tondeurs having dinner in Sendai, Japan in 2005.

achievements. Mathematics flourishes as a truly global intellectual enterprise with fantastic connectivity and universal standards, as demonstrated by the holding of periodic worldwide International Congresses of Mathematicians. I fervently hope that the United States will continue to be an exemplary steward of mathematics and science—together with other nations.

MP: *Do you have any closing words?*

Tondeur: In retirement, in the company of Claire, my cherished wife, we have deepened our knowledge of history and civilization through study and travel, getting an ever more profound sense of mankind's evolutionary dynamics within "Big History." We frequently are exhilarated by the beauty of nature's evolution and man's creations and are sometimes disheartened by man's destructiveness and folly. It is a privilege to be able to live mainly for the letters and sciences, which continues to bring us into the company of some of mankind's most gifted members (past and present). For me, having mathematics play a central and increasingly unifying role in the sciences is something that I am particularly grateful to have been able to devote heart and soul to it during an entire lifetime.

Biographical Notes

Lars Valerian Ahlfors—Born in Helsingfors, Finland, April 18, 1907. PhD, 1928, Helsinki University. Postdoctoral appointments: Eidgenösische Technische Hochschule, Zürich; Aaboe Academy, Turku. Positions held: Harvard University, 1935–38; University of Helsinki, 1938–44; University of Zürich, 1945–46, 1977; Harvard University, 1946–77. Honors: Fields Medal, 1936; Wolf Prize, 1981; Leroy P. Steele Prize, AMS, 1982. Died in Pittsfield, MA, October 11, 1996.

Tom M. Apostol—Born in Helper, UT, August 20, 1923. BS, 1944, MS, 1946, University of Washington; PhD, 1948, University of California, Berkeley. Positions held: University of California, Berkeley, 1948–49; Massachusetts Institute of Technology, 1949–50; California Institute of Technology, 1950–present. Honors: Lester R. Ford Award, MAA, 2005, 2008.

Harold Maile Bacon—Born in Los Angeles, CA, January 13, 1907. BA, 1928, MS, 1929, PhD, 1933, Stanford University. Positions held: Stanford University 1933–72. Honors: Dinkelspiel Award, Stanford University, 1965; Distinguished Service Award, MAA, 1988. Died in Palo Alto, CA, August 22, 1992.

Thomas Francis Banchoff—Born in Trenton, NJ, April 7, 1938. BA, 1960, University of Notre Dame; MS, 1962, PhD, 1964, University of California, Berkeley. Positions held: Harvard University, 1964–66; University of Amsterdam, 1966–67; Brown University, 1967–present. Honors: Lester R. Ford Award, MAA, 1978; Bray Award, Brown University, 1993; Haimo Award for Distinguished Teaching, MAA, 1996; Director's Award for Distinguished Teacher Scholar, NSF, 2004.

Leon Bankoff—Born in New York, NY, December 13, 1908. BS, 1928, City College of New York; DDS, 1932, University of Southern California. A dentist by profession, he practiced in Beverly Hills. Died in Los Angeles, CA, February 16, 1997.

Alice Beckenbach—Born December 16, 1917. BS, 1938, Northwestern University. Died in Hamilton, NY, March 18, 2010.

Arthur Benjamin—Born in Mayfield Heights, OH, March 19, 1961. BS, 1983, Carnegie Mellon University; PhD, 1989, The Johns Hopkins University. Positions held: Harvey Mudd College, 1989–present. Honors: Haimo Award for Distinguished Teaching, MAA, 2000; George Pólya Lecturer, MAA, 2006-8; Beckenbach Book Prize, MAA, 2006.

Dame Mary Lucy Cartwright—Born in Aynho, Northamptonshire, England, December 17, 1900. Undergraduate, 1923, St. Hugh's College, Oxford University; MA, Oxford University; MA, Cambridge University; DPhil, 1930, Oxford University. Positions held: Alice Ottley School, 1923–27; Wycombe School, 1927–28; Cambridge University, 1933–98. Honors: Fellow of the Royal Society, 1947; Sylvester Medal, RS, 1964; De Morgan Medal, LMS, 1968; Dame Commander of the Order of the British Empire, 1969. Died in Cambridge, England, April 3, 1998.

Joseph Gallian—Born in New Kensington, PA, January 5, 1942. BA, 1966, Slippery Rock University; MA, 1968, University of Kansas; PhD, 1971, University of Notre Dame. Positions held: University of Notre Dame, 1971–72; University of Minnesota, Duluth, 1972–present. Honors: Carl B. Allendoerfer Award, MAA, 1977; Haimo Distinguished Teaching Award, MAA, 1993; Trevor Evans Award, MAA, 1996; George Pólya Lecturer, MAA, 1999–2001.

Abbreviations: AAAS, American Association for the Advancement of Science; AMS, American Mathematical Society; APS, American Philosophical Society; AWM, Association for Women in Mathematics; AWIS, Association for Women in Science; LMS, London Mathematical Society, MAA, Mathematical Association of America; NSA, National Security Agency; NSF, National Science Foundation; RS, Royal Society of London.

Richard Kenneth Guy—Born in Nuneaton, Warwickshire, England, September 30, 1916. BA, 1938, MA, 1941, Cambridge University. Positions held: Goldsmiths (University of London), 1947–51; University of Malay, Singapore, 1951–62; Indian Institute of Technology, New Delhi, 1962–65; University of Calgary, 1965–present. Honors: Fellow, Cambridge Philosophical Society, 1976; Hedrick Lecturer, MAA, 1978; Lester R. Ford Award, MAA, 1989.

Fern Hunt—Born in New York, NY, 1948. AB, Bryn Mawr College; MS, PhD, 1978, Courant Institute of Mathematical Sciences, New York University. Postdoctoral appointment, University of Utah, 1978. Positions held: Howard University, 1978–93; Consultant, National Bureau of Standards, 1986–91; National Institutes of Health, 1981–82; National Institute of Standards and Technology, 1993–present. Honor: Arthur S. Flemming Award, 2000.

Dusa Waddington McDuff—Born in London, October 18, 1945. BSc, 1967, University of Edinburgh; PhD, 1971, Cambridge University. Postdoctoral appointments: Cambridge University, Moscow University, 1971–73. Positions held: York University, 1973–76; Massachusetts Institute of Technology, 1974–75; Warwick University, 1976–78; State University of New York, Stony Brook, 1978–2007; Columbia University, 2007–present. Honors: Ruth Lyttle Satter Prize, AMS, 1991; Fellow of the Royal Society, 1994; American Academy of Arts and Sciences, 1995; Noether Lecturer, AWM, 1998; National Academy of Sciences (U.S.), 1999; Honorary membership, London Mathematical Society, 2007.

Donald G. Saari—Born in Houghton, MI, March 9, 1940. BS, Michigan Technological University; MS, 1962, PhD, 1964, Purdue University. Postdoctoral appointment, Yale University, 1967–68. Positions held: Northwestern University, 1968–2000; University of California, Irvine, 2000–present. Honors: Lester R. Ford Award, MAA, 1985; Guggenheim Fellow, 1989; Chauvenet Prize, MAA, 1995; Carl B. Allendoerfer Award, MAA, 1999; National Academy of Sciences (U.S.), 2001; American Academy of Arts and Sciences, 2004.

Atle Selberg—Born in Langeslund, Norway, June 14, 1917. PhD, 1943, University of Oslo. Positions held: University of Oslo, 1942–47; Institute for Advanced Study, Princeton, 1947–48; Syracuse University, 1948–49; Institute for Advanced Study, Princeton, 1949–2007. Honors: Fields Medal, 1950; Wolf Prize, 1986; Royal Norwegian Order of St. Olav, 1987. Died in Princeton, NJ, August 6, 2007.

Jean Ellen Taylor—Born in San Mateo, CA, September 17, 1944. BA, 1966, Mount Holyoke College; MS, 1968, University of California, Berkeley; MS, 1971, University of Warwick; PhD, 1973, Princeton University. Positions held: Massachusetts Institute of Technology, 1972–73; Rutgers University, 1973–2002; Visiting Scholar, Courant Institute of Mathematical Sciences, New York University, 2002–present. Honors: Earl Raymond Hedrick Lecturer, MAA, 1998; Fellow, AAAS, 1999; Fellow, American Academy of Arts and Sciences, 1999; Fellow, AWIS, 1999; and Noether Lecturer, AWM, 2003.

Philippe Tondeur—Born in Zürich, Switzerland, December 7, 1932. PhD, 1961, University of Zürich. Positions held: University of Paris, 1961–63; University of Zürich, 1963–64; Harvard University, 1964–65; University of California, Berkeley, 1965–66; Wesleyan University, 1966–68; University of Illinois, Champaign-Urbana, 1970–2002; Director, Division of Mathematical Sciences, National Science Foundation, 1999–2002.

Glossary

We add here some definitions (often truncated and simplified) of some mathematical names, words, and phrases alluded to in the interviews. (We beg the indulgence of professional mathematicians who may find them painfully inadequate.) For some of the more technical terms and theorems cited, please refer to google.com or, more specifically, to mathworld.wolfram.com or to planetmath.org, since to describe them here would involve far too many additional definitions to make them useful to most readers.

Algebraic geometry: The study of the geometry of curves, surfaces, and higher-dimensional objects defined by systems of polynomial equations.

Algebraic topology: The study of topological properties using methods from abstract algebra.

Analytic function: An analytic (holomorphic) function is one that can be expressed in terms of all of its higher-order derivatives.

Analytic number theory: The study of the whole numbers using methods and results from calculus and the theory of functions of a complex variable.

Archimedean solids: The thirteen semiregular solids, originally discovered by Archimedes, whose faces are all regular polygons and whose vertices are all surrounded by the same arrangement of faces.

Bourbaki: Collective pen name of a group of (mainly French) mathematicians who set out to write a formal and rigorous account of "elementary" mathematics. Bourbaki also runs a seminar that meets three times a year in Paris to discuss recent mathematical developments.

Brauer-Fowler theorem: A crucial step in the long project of classifying the finite simple groups, those that have no nontrivial normal subgroups.

Cauchy-Bunyakovskii-Schwarz inequality: If a_i, $b_i \geq 0$, $i = 1, 2, \ldots, n$, then
$$(a_1 b_1 + \ldots + a_n b_n)^2 \leq (a_1^2 + \ldots + a_n^2)(b_1^2 + \ldots + b_n^2).$$

Cayley graphs: A pictorial representation of the structure of a group.

Chebyshev inequality: An inequality in probability theory that states that in any probability distribution "almost all" of the values are "close to the mean."

Combinatorics: The branch of mathematics concerned with counting, often making use of methods from calculus and algebra.

Complex function theory: The exploration of properties of functions where the variables are no longer real numbers but complex numbers, those of the form $a + bi$, where a and b are real and $i^2 = -1$.

Composite number: A positive whole number that is not a prime.

Conformal mappings: Transformations that preserve the size and the sign of angles and hence the overall shape of a geometric figure.

Dedekind cut: A technical device used to define irrational numbers in terms of rational numbers.

Dense set (of real numbers): A set of numbers sufficiently large to provide arbitrarily good approximations for all other numbers.

Differential geometry: The study of curves and surfaces using techniques from differential calculus.

Dirichlet series: A series of the form
$$a_1/1^s + a_2/2^s + a_3/3^s + \ldots.$$
The Riemann zeta function and its generalizations are examples of functions defined by Dirichlet series.

Dynamical systems: Any system that evolves over time, particularly if described by differential equations.

Entire function: A function of a complex variable that is holomorphic at every point in the finite complex plane.

Erdős-Mordell theorem (sometimes referred to as the Erdős-Mordell inequality): For a point P inside a triangle ABC, the sum of the three distances from P to the vertices is greater than or equal to twice the sum of the distances from P to the three sides.

Erdős numbers: An assignment of numbers to mathematicians that goes as follows: if a person has coauthored a paper with Paul Erdős the number assigned to that person is 1; if another person (who has not coauthored a paper with Erdős) has coauthored a paper with a person having Erdős number 1, then the second person has Erdős number 2; and so on. It sets up a partial ordering of mathematicians that depends on their "distances" from Erdős.

Euler's formula: $e^{i\theta} = \cos\theta + i\sin\theta$, where $i^2 = -1$; a formula connecting the exponential function with trigonometric functions.

Fermat's equation ("last theorem"): A famous conjecture for which Fermat claimed a proof in 1670, but with no details provided: there are no x, y, and z, positive whole numbers, that satisfy the equation $x^n + y^n = z^n$, for n greater than 2. Finally, Andrew J. Wiles provided a proof in 1994.

Field: A set of numbers for which addition, subtraction, multiplication, and division work as they do for the real numbers.

Fields Medal: Widely viewed as the highest honor awarded to mathematicians, this prize was initiated with funds from the International Congress of Mathematicians in Toronto in 1924 and named for the president of that Congress, John Charles Fields. It was first awarded in 1936 at the Congress in Oslo, where one of the first two recipients was Lars Ahlfors.

Floer theory: A part of differential geometry with implications in symplectic geometry, topology, and mathematical physics.

Functional analysis: The study of linear spaces, especially linear spaces of functions.

Game theory: The mathematical theory involved in the choice of strategy in decision making.

Geometric function theory: The study of the geometric properties of analytic functions. A principal result in this area is the Riemann mapping theorem that says that under certain conditions there is a conformal function that maps a region in the complex plane into a disk.

Graph theory: The study of the properties of networks formed by choosing a set of vertices and connecting some of them by edges.

Group: An algebraic structure consisting of a set with a single operation, often used to study problems involving symmetry.

Hamiltonian circuit: In a (connected) graph, a Hamiltonian circuit is a path (a sequence of edges) that passes in turn through each of the points (vertices) once and returns to the starting point. The problem of finding the shortest possible Hamiltonian circuit is often called the "Travelling Salesman Problem" (still unsolved) that would allow a salesman to visit each of his customers in a way involving the least travel distance or, in other words, the least time to make his calls.

Hypercube: The analogue of a cube in spaces of dimension higher than three.

Kleinian groups: A class of groups that arise in the study of conformal (angle-preserving) transformations of the unit ball in 3-space.

Magic square: A square array of integers (whole numbers) in which the sums of the integers in each row, each column and each diagonal are all the same.

Modern algebra: Often called "abstract algebra," this is the study of algebraic structures such as groups, rings, fields, vector spaces, and modules, among others.

Nachlass: A legacy, in particular, in the case of a scholar or scientist, the collection of notes, manuscripts and correspondence related to the person's work.

Number theory: The branch of mathematics involving properties of integers (whole numbers).

Ordinary differential equation: An equation involving a function of one variable and its derivatives, the solution of which is a function or a set of functions. In many examples, the variable is time, yielding a dynamical system.

Peano axioms: A small set of defining properties that characterize the natural numbers: 0, 1, 2, 3, ...

Pell's equation: The problem of finding integers x and y such that $x^2 - ny^2 = 1$ where n is an integer (a whole number) that is not a square.

Point set topology: Sometimes called general topology, it is the study of those properties of sets of points and functions between them, especially continuity, dimension, and connectedness.

Pólya (enumeration) theorem: Sometimes named for Federov—who discovered it first—or Redfield—who published it first but after which it remained unknown—it was promulgated by George Pólya. This theorem allows one, for example, to count colorings of constituent parts of polyhedra that appear to be different but are in fact the same, just seen from different viewpoints. Commonly applied in chemistry in the study of molecular structures.

Polyominoes: A generalization of the shape of a domino but with larger numbers of squares placed together to form more elaborate tiles; a concept from recreational mathematics.

Prime: A prime, p, is a positive whole number ≥ 2 that has only 1 and p as factors.

Prime Number Theorem: If $\pi(x)$ is the number of primes up to and including x, then $\pi(x)$ is asymptotically (that is, for very large x) $x/\ln x$.

Quasiconformal mappings: In complex variables these are mappings that are "almost" conformal, that is, angle distortion is limited within certain technical constraints.

Riemann hypothesis: A conjecture by Bernhard Riemann in 1857 that the (nontrivial) zeros (points in the complex plane where the Riemann zeta function takes on the value 0) lie on a vertical line in the plane where the real part is equal to ½. The conjecture remains unproved and is widely viewed as the most important and tantalizing unsolved problem in pure mathematics.

Riemann zeta function: The Dirichlet series where $a_i = 1$ for $i = 1, 2, 3, \ldots$ The zeta-function "encodes" many properties of the prime numbers. (See Dirichlet series.)

Ring: An algebraic structure with two operations motivated by the study of the integers or sets of polynomials.

Selberg sieve: A generalization of Eratosthenes's sieve, a method in analytic number theory for locating primes among the positive integers.

String theory: An active area of research in particle physics that is concerned with reconciling quantum mechanics and general relativity. The field is often cited in referring to the work of Edward Witten, the only physicist to have received the Fields Medal in mathematics.

Summability: A field of mathematics concerned with techniques assigning values to divergent series.

Symplectic geometry: A branch of differential geometry or differential topology, with origins in classical mechanics.

Tauberian theorems: A family of theorems that serve as crucial tools in analytic number theory and complex analysis.

Theory of numbers: (See Number theory.)

Topology: The study of properties of geometric figures that are unchanged under certain transformations; sometimes called "rubber sheet geometry."

Uniform convergence: A property of a set of functions on a given set that converge at the same rate for all members of the set.

Van der Pol's equation: The ordinary differential equation

$$u'' + \alpha(u^2 - 1)u' + \beta u = 0$$

that has one and only one periodic solution.

Vector geometry: In n-dimensional space, the study of geometry using vectors, objects that, unlike numbers, have not only magnitude but direction as well.

Voting theory: The study of various systems for setting up elections, where the choice of the system can affect the outcome. One of the principal and most surprising results is the Arrow impossibility theorem, sometimes called Arrow's paradox, which says, roughly, that no "fair" system can be devised that satisfies three conditions, each of which seems entirely reasonable to expect of any voting system. Kenneth Arrow won the Nobel Prize in Economics in 1972 for related work.

Index